STATISTICS
USING
PYTHON

STATISTICS
USING
PYTHON

Oswald Campesato

MERCURY LEARNING AND INFORMATION
Boston, Massachusetts

Publisher: David Pallai
MERCURY LEARNING AND INFORMATION
121 High Street, 3rd Floor
Boston, MA 02110
info@merclearning.com
www.merclearning.com
800-232-0223

O. Campesato. *Statistics Using Python.*
ISBN: 9781683928805

Library of Congress Control Number: 2023918207

232425321 This book is printed on acid-free paper in the United States of America.

Our titles are available for adoption, license, or bulk purchase by institutions, corporations, etc. For additional information, please contact the Customer Service Dept. at 800-232-0223(toll free).

All of our titles are available in digital format at academiccourseware.com and other digital vendors. *Companion files (code listings) for this title are available by contacting* info@merclearning.com. The sole obligation of MERCURY LEARNING AND INFORMATION to the purchaser is to replace the disc, based on defective materials or faulty workmanship, but not based on the operation or functionality of the product.

I'd like to dedicate this book to my parents
– may this bring joy and happiness into their lives

CONTENTS

Preface   *xi*

CHAPTER 1:   **Working with Data**   **1**
What is Data Literacy?   1
Exploratory Data Analysis (EDA)   2
Dealing with Data: What Can Go Wrong?   5
An Explanation of Data Types   7
Working with Data Types   12
What is Drift?   13
Discrete Data Versus Continuous Data   14
Binning Data Values   15
Correlation   17
Working with Synthetic Data   19
Summary   27

CHAPTER 2:   **Introduction to Probability**   **29**
What is Set Theory?   29
Open, Closed, Compact, and Convex Sets (Optional)   31
Concepts in Probability   32
Set Theory and Probability   36
Coin Tossing Probabilities   37
Dice Tossing Probabilities   43
Card Drawing Probabilities   47

	Container-Based Probabilities	49
	Children-Related Probabilities	52
	Summary	53
CHAPTER 3:	**Introduction to Statistics**	**55**
	Introduction to Statistics	55
	Basic Concepts in Statistics	56
	The Variance and Standard Deviation	62
	The Moments of a Function (Optional)	64
	Random Variables	67
	Multiple Random Variables	69
	Sampling Techniques for a Population	70
	What is Bias?	72
	Two Important Results in Probability	73
	Summary	74
CHAPTER 4:	**Metrics in Statistics**	**75**
	The Confusion Matrix	75
	The ROC Curve and AUC Curve	87
	The sklearn.metrics Module (Optional)	92
	Statistical Metrics for Categorical Data	93
	Metrics for Continuous Data	97
	MAE, MSE, and RMSE	100
	Approximating Linear Data with np.linspace()	102
	Summary	103
CHAPTER 5:	**Probability Distributions**	**105**
	PDF, CDF, and PMF	105
	Two Types of Probability Distributions	107
	Discrete Probability Distributions	108
	Continuous Probability Distributions	116
	Advanced Probability Functions	136
	Non-Gaussian Distributions	140

The Best-Fitting Distribution for Data   140

Summary   148

CHAPTER 6:   **Hypothesis Testing**   **149**

What is Hypothesis Testing?   149

Components of Hypothesis Testing   151

Test Statistics   152

Working with p-values   154

Working with Alpha Values   157

Point Estimation, Confidence Level, and Confidence Intervals   157

What is A/B Testing?   161

The Lifespan of an A/B Test   165

Maximum Likelihood Estimation (MLE)   166

Summary   168

Appendix A: **Introduction to Python**   **169**

Tools for Python   169

Python Installation   171

Setting the PATH Environment Variable (Windows Only)   172

Launching Python on Your Machine   172

Identifiers   174

Lines, Indentation, and Multi-Line Statements   174

Quotation Marks and Comments   175

Saving Your Code in a Module   177

Some Standard Modules   178

The help() and dir() Functions   178

Compile Time and Runtime Code Checking   180

Simple Data Types   180

Working with Numbers   181

Working with Fractions   185

Unicode and UTF-8   186

	Working with Strings	187
	Slicing and Splicing Strings	190
	Search and Replace a String in Other Strings	192
	Remove Leading and Trailing Characters	193
	Printing Text without New Line Characters	194
	Text Alignment	195
	Working with Dates	196
	Exception Handling	198
	Handling User Input	200
	Python and Emojis (Optional)	203
	Command-Line Arguments	204
	Summary	205
Appendix B:	**Introduction to Pandas**	**207**
	What is Pandas?	207
	A Pandas Data Frame with a NumPy Example	210
	Describing a Pandas Data Frame	213
	Boolean Data Frames	216
	Data Frames and Random Numbers	218
	Reading CSV Files in Pandas	220
	The loc() and iloc() Methods	222
	Converting Categorical Data to Numeric Data	222
	Matching and Splitting Strings	227
	Converting Strings to Dates	230
	Working with Date Ranges	232
	Detecting Missing Dates	234
	Interpolating Missing Dates	236
	Other Operations with Dates	239
	Merging and Splitting Columns in Pandas	244
	Reading HTML Web Pages	247
	Saving a Pandas Data Frame as an HTML Web Page	248
	Summary	251
Index		253

PREFACE

Statistics Using Python is your definitive guide to understanding statistical concepts through Python programming. In the current digital age, data plays an integral role in business and decision-making. The ability to extract meaningful insights from data requires a deep understanding of statistics. This book provides a comprehensive and practical approach to statistics using Python, one of the most popular and versatile programming languages in the world today.

Statistics Using Python is designed to offer a fast-paced yet thorough introduction to essential statistical concepts using Python code samples, aiming to assist data scientists in their daily endeavors. While the book casts a wide net to cater to a broad audience, it ensures that each topic is introduced with clarity, followed by executable Python code samples that can be modified and applied according to individual needs.

Target Audience: This book primarily targets data scientists and enthusiasts who have a foundational understanding of statistics but wish to delve deeper. Whether you are a beginner wanting to grasp the basics or someone with intermediate knowledge aiming to broaden your statistical horizon, this book offers a structured approach to various concepts. The interleaving of foundational and advanced topics ensures readers can pace their learning according to their comfort and familiarity.

KEY TOPICS COVERED

- Working with Data: Data preprocessing, cleaning, and exploratory analysis.
- Basics of Probability: Understand the foundation upon which statistical concepts are built.
- Introduction to Statistics: Learn about descriptive and inferential statistics and their applications.
- Metrics in Statistics: Gauge the accuracy, precision, and other metrics vital in data analysis.
- Probability Distributions: Explore various distributions and their real-world significance.
- Hypothesis Testing: Learn to make informed decisions from data.

FEATURES

- **Hands-On Approach**: Each topic is accompanied by Python code samples to ensure readers can apply what they learn immediately.
- **Wide Range of Topics**: From basic data handling to advanced statistical concepts, this book provides a comprehensive coverage.
- **Supplementary Material**: Downloadable code samples and figures from the publisher ensure you have all the tools at your disposal.
- **Two Appendices:** An *Introduction to Python* and an *Introduction to Pandas* provide refresher material if needed.

Enjoy the world of statistics using Python and understand the power and usefulness of data in the digital age.

O. Campesato
December 2023

## CHAPTER 1

WORKING WITH DATA

This chapter focuses on data types that you will encounter in datasets, including currency and dates, as well as scaling data values.

The first part of this chapter briefly discusses some aspects of Exploratory Data Analysis (EDA), such as data quality and data-centric AI versus model-centric AI, as well as some of the steps involved in data cleaning and data wrangling. You will also see an EDA code sample involving the Titanic dataset.

The second part of this chapter describes common types of data, such as binary, nominal, ordinal, and categorical data. In addition, you will learn about continuous versus discrete data, quantitative and quantitative data, and types of statistical data.

The third second introduces the notion of data drift and data leakage, followed by model selection. This section also describes how to process categorical data, and how to map categorical data to numeric data.

WHAT IS DATA LITERACY?

There are various definitions of data literacy that involve concepts such as data, meaningful information, decision-making, drawing conclusions, and chart reading. According to Wikipedia, data literacy is defined as follows:

> "Data literacy is the ability to read, understand, create, and communicate data as information. Much like literacy as a general concept, data literacy focuses on the competencies involved in working with data. It is, however, not similar to the ability to read text since it requires certain skills involving reading and understanding data." (*https://en.wikipedia.org/wiki/Data_literacy*)

Data literacy encompasses many topics, starting with analyzing data that is often in the form of a comma separated values (CSV) file. The quality of the data in a dataset is of paramount importance: high data quality enables you to make more reliable inferences regarding the nature of the data. Indeed, high data quality is a requirement for fields such as machine learning and scientific experiments. However, you might face various challenges regarding robust data, such as

- a limited amount of available data
- costly acquisition of relevant data
- difficulty in generating valid synthetic data
- availability of domain experts

Depending on the domain, the cost of data cleaning can involve months of work at a cost of millions of dollars. For instance, identifying images of cats and dogs is essentially trivial, whereas identifying potential tumors in x-rays is much more costly and requires highly skilled individuals.

With all the preceding points in mind, let's take a look at EDA, which is the topic of the next section.

EXPLORATORY DATA ANALYSIS (EDA)

According to Wikipedia, EDA involves analyzing datasets to summarize their main characteristics, often with visual methods. EDA also involves searching through data to detect patterns (if there are any) and anomalies, and in some cases, test hypotheses regarding the distribution of the data.

EDA represents the initial phase of data analysis, whereby data is explored to determine its primary characteristics. Moreover, this phase involves detecting patterns (if any), and any outstanding issues pertaining to the data. The purpose of EDA is to obtain an understanding of the semantics of the data without performing a deep assessment of the nature of the data. The analysis is often performed through data visualization to produce a summary of their most important characteristics. The four types of EDA are listed here:

- univariate non-graphical
- multivariate non-graphical
- univariate graphical
- multivariate graphical

In brief, the two primary methods for data analysis are *qualitative data analysis* techniques and *quantitative data analysis* techniques.

As an example of exploratory data analysis, consider the plethora of cell phones that customers can purchase for various needs (such as for work, home, and minors). Visualize the data in an associated dataset to determine the top ten (or top three) most popular cell phones, which can potentially be performed by state (or province) and country.

An example of quantitative data analysis involves measuring (quantifying) data, which can be gathered from physical devices, surveys, or activities such as downloading applications from a Web page.

Common visualization techniques used in EDA include histograms, line graphs, bar charts, box plots, and multivariate charts.

What is Data Quality?

According to Wikipedia, *data quality* refers to "the state of qualitative or quantitative pieces of information." (*https://en.wikipedia.org/wiki/Data_quality*) Furthermore, high data quality refers to data whose quality meets the various needs of an organization. In particular, performing data cleaning tasks are the type of tasks that assist in achieving high data quality.

When companies label their data, they obviously strive for a high quality of labeled data, and yet the quality can be adversely affected in various ways, some of which are listed here:

- inaccurate methodology for labeling data
- insufficient data accuracy
- insufficient attention to data management

The cumulative effect of the preceding (and other) types of errors can be significant, to the extent that models underperform in a production environment. In addition to the technical aspects, underperforming models can have an adverse effect on business revenue.

Related to data quality is *data quality assurance*, which typically involves data cleaning tasks, after which data is analyzed to detect potential inconsistencies in the data, and then determine how to resolve those inconsistencies. Another aspect to consider: the aggregation of additional data sources, especially involving heterogenous sources of data, can introduce challenges with respect to ensuring data quality. Other concepts related to data quality include *data stewardship* and *data governance*, both of which are discussed in multiple online articles.

Data-Centric AI or Model-Centric AI?

A *model-centric* approach focuses primarily on enhancing the performance of a given model, and data is considered secondary in importance. In fact, during the past ten years or so, the emphasis of AI has been a model-centric approach. Note that during this time span, some very powerful models and architectures have been developed, such as the CNN model for image classification in 2012 and the enormous impact (especially in NLP) of models based on the transformer architecture that was developed in 2017.

By contrast, a *data-centric* approach concentrates on improving data, which relies on several factors, such as the quality of labels for the data as well as obtaining accurate data for training a model.

Given the importance of high-quality data with respect to training a model, it stands to reason that using a data-centric approach instead of a model-centric approach can result in higher quality models in AI. While data quality and model effectiveness are both important, keep in mind that the data-centric approach is becoming increasingly more strategic in the machine learning world. There is more information available online:

https://research.aimultiple.com/data-centric-ai/

The Data Cleaning and Data Wrangling Steps

The next step often involves *data cleaning* to find and correct errors in the dataset, such as missing data, duplicate data, or invalid data. This task also involves data consistency, which pertains to updating different representations of the same value in a consistent manner. As a simple example, suppose that a Web page contains a form with an input field whose valid input is either Y or N, but users are able to enter Yes, Ys, or ys as text input. Obviously, these values correspond to the value Y, and they must all be converted to the value Y to achieve data consistency.

Data wrangling can be performed after the data cleaning task is completed. Although interpretations of data wrangling do vary, in this book, the term refers to transforming datasets into different formats as well as combining two or more datasets. Hence, data wrangling does not examine the individual data values to determine whether they are valid: this step is performed during data cleaning.

Sometimes, it is worthwhile to perform another data cleaning step after the data wrangling step. For example, suppose that two CSV files contain employee-related data, and you merge these CSV files into a third CSV file.

The newly created CSV file might contain duplicate values: it is certainly possible to have two people with the same name (such as John Smith).

ELT and ETL

Extract, Load, and Transform (ELT) is a pipeline-based approach for managing data. Another pipeline-based approach is Extract, Transform, and Load (ETL), which is more popular than ELT. However, ELT has the following advantages over ETL:

- ELT requires less computational time.
- ELT is well-suited for processing large datasets.
- ELT is more cost-effective.

ELT involves 1) data extraction from one or more sources, 2) transforming the raw data into a suitable format, and 3) loading the result into a data warehouse. The data in the warehouse becomes available for additional analysis.

DEALING WITH DATA: WHAT CAN GO WRONG?

In a perfect world, all datasets are in pristine condition, with no extreme values, no missing values, and no erroneous values. Every feature value is captured correctly, with no chance for any confusion. Moreover, no conversion is required between date formats, currency values, or languages because of the "One Universal Standard" that defines the correct formats and acceptable values for every possible set of data values.

Of course, all the problems mentioned in the previous paragraph can and do occur, which is the reason for the techniques that are discussed in this chapter. Even after you manage to create a wonderfully clean and robust dataset, other issues can arise, such as data drift.

In fact, the task of cleaning data is not necessarily complete even after a machine learning model is deployed to a production environment. For instance, an online system that gathers terabytes or petabytes of data on a daily basis can contain skewed values that adversely affect the performance of the model. Such adverse effects can be revealed through the changes in the metrics that are associated with the production model.

Datasets

In simple terms, a *dataset* is a source of data (such as a text file) that contains rows and columns of data. Each row is typically called a "data point," and each column is called a "feature." A dataset can be a CSV or tab separated values (TSV) file, Excel spreadsheet, a table in a Relational Database Management System (RDBMS), a document in a NoSQL database, the output from a Web Service, and so forth.

Note that a *static dataset* consists of fixed data. For example, a CSV file that contains the states of the USA is a static dataset. A slightly different example involves a product table that contains information about the products that customers can buy from a company. Such a table is static if no new products are added to the table. Discontinued products are probably maintained as historical data that can appear in product-related reports.

By contrast, a *dynamic dataset* consists of data that changes over a period of time. Simple examples include housing prices, stock prices, and time-based data from Internet of Things (IoT) devices.

A dataset can vary from very small (perhaps a few features and 100 rows) to very large (more than 1,000 features and more than one million rows). If you are unfamiliar with the problem domain for a particular dataset, then you might struggle to determine its most important features. In this situation, you consult a domain expert who understands the importance of the features, their inter-dependencies (if any), and whether the data values for the features are valid. In addition, there are algorithms (called dimensionality reduction algorithms) that can help you determine the most important features, such as Principal Component Analysis (PCA).

Before delving into topics such as data preprocessing and data types, let's take a brief detour to introduce the concept of feature importance, which is the topic of the next section.

As you will see, someone needs to analyze the dataset to determine which features are the most important and which features can be safely ignored to train a model with the given dataset. A dataset can contain various data types, such as

- Audio data
- Image data
- Numeric data
- Text-based data
- Video data
- Combinations of the above

In this book, we will only consider datasets that contain columns with numeric or text-based data types, which can be further classified as follows:

- nominal (string-based or numeric)
- ordinal (ordered values)
- categorical (enumeration)
- interval (positive/negative values)
- ratio (non-negative values)

The next section contains brief descriptions of the data types that are in the preceding list.

AN EXPLANATION OF DATA TYPES

This section contains brief descriptions of the following data types:

- binary data
- nominal data
- ordinal data
- categorical data
- interval data
- ratio data

Later, you will learn about the difference between continuous data versus discrete data, as well as the difference between qualitative data versus quantitative data. In addition, the Pandas documentation describes data types and how to use them in Python.

Binary Data

Binary-valued data involves data that can only take two distinct values. As such, binary data is the simplest type of data. A common example involves flipping a coin: the only outcomes are in the set {H, T}. Other terms for binary data include dichotomous, logical data, Boolean data, and indicator data. Binary data is also a type of categorical data that is discussed later.

Nominal Data

The word "nominal" has multiple meanings, and in this book, it refers to something that constitutes a name (the prefix "nom" means "name"). Thus, *nominal data* is often name-based data that involves different name labels.

Examples of nominal data include hair color, music preferences, and movie types. As you can see, there is no hierarchy or ordering involved, so all values have the same importance. However, the number of items in nominal values might be different, such as the number of people that belong to different political parties.

Nominal data can involve numeric values to represent different values of a feature. For example, the numbers in the set {0,1} can represent {Male, Female}, and the numbers in the set {0,1,2,3,4,5,6} can represent the days in a week. However, there is no hierarchical interpretation associated with these numeric values: the day of the week represented by "0" is not considered more important or more valuable than any of the other numeric labels for the other days of the week. Instead, think of each element in terms of its predecessor and successor: note that the first element has no predecessor and the last element has no successor. If you are familiar with programming languages, the counterpart to integer-based nominal values would be an enumeration, an example of which is here:

```
enum DAY {SUN,MON,TUE,WED,THU,FRI,SAT};
```

Ordinal Data

Ordinal data implies an ordering of the elements in a finite set (think "ordering" from the prefix "ord"). For example, there are different values for titles regarding software developers. As a simplified example, the set consisting of {D1, D2, SD1, SD2} can be used to specify junior developers (D1) up through senior developers (SD2), which have criteria associated with each level. Hence, integer-based and string-based elements of ordinal data are ordered.

Keep in mind that integer-based ordinal data does not have an implied relative value. For example, consider the following set of ordinal data S = {1,2,3,4,5,6} that represents grade levels in an organization. A level 2 employee is not "twice" as experienced as a level 1 employee, nor would a level 6 employee be three times as experienced as a level 2 (unless you define these values in such a manner).

Please read the scikit-learn documentation regarding the class OrdinalEncoder (scikit-learn.preprocessing.OrdinalEncorder) for handling ordinal data. More information about ordinal data can be found online:

https://careerfoundry.com/en/blog/data-analytics/what-is-ordinal-data/

Categorical Data

Categorical data refers to nominal data as well as ordinal data: please read the preceding sections regarding the nuances involved in nominal data and ordinal data. Categorical data can only assume a finite set of distinct values, such as enumerations that are explained in a previous section. In addition, Pandas can explicitly specify a column as type "categorical" when you read the contents of a CSV file via the `read_csv()` method.

Interval Data

Interval data pertains to data that is ordered and lies in an interval or range, such as the integers and floating-point numbers in the interval [-1,1]. Examples of interval data include temperature and income-versus-debt. As you can see, interval data values can be negative as well as positive.

Ratio Data

Ratio data involves measured intervals, such as barometric pressure, height, altitude, and so forth. Notice the difference between interval data and ratio data: unlike interval data, ratio data *cannot* be negative. Obviously, it makes no sense to refer to negative values for the barometric pressure, a person's height, or altitude above the surface of the earth.

Continuous Data Versus Discrete Data

Continuous data can take on any value in an interval, such as [-1,1], [0,1], or [5,10]. Hence, continuous data involves floating point numbers, which includes interval data. An *interval* contains an uncountably infinite number of values.

One other point to note pertains to possible values and their floating-point representation. For instance, a random number in the interval [0,1] involves an uncountably infinite number of values, whereas its representation as a floating-point number is limited to a large yet finite number of values. Let's suppose that the integer 10^1000 equals the number of numbers in the interval [0,1] that can be represented as a floating-point number. Then the smallest positive number in the interval [0,1] that can be represented as a floating-point number is 1/N. However, there is an uncountably infinite number of values in the interval [0,1/N), which we would approximate as the value 0 (or possibly 1/N).

Discrete data can take on a finite set of values, and the earlier comments regarding successors and predecessors apply to discrete data. As a simple example, the outcome of tossing a coin or throwing a die (or multiple dice) involves discrete data, which are also examples of nominal data. In addition, the associated probabilities for the outcomes form a discrete probability distribution (discussed later).

Qualitative and Quantitative Data

Quantitative data can be either discrete or continuous. For example, a person's age that is measured in years is discrete, whereas the height of a person is continuous. The word "continuous" in statistics does not always have the same meaning when it is used in a mathematical context. For instance, the price of a house is treated as a continuous feature, but it is not continuous in the mathematical sense. This is because the smallest unit of measure is a penny, and there are many (in fact, an uncountably infinite number of) values between two consecutive penny values. Here are two examples of discrete data values, followed by three examples of continuous data values:

- revenue (money)
- number of items sold
- water temperature
- wind speed
- vehicle velocity

Each of the preceding data values are numeric types involving something that has a business impact or physical characteristic.

Qualitative data can sometimes be represented as string-based values, such as different types of color or movie genres. Hence, nominal data and ordinal data are considered qualitative data.

As you saw in an earlier section, it is possible to use integer-based values for nominal values, such as days of the week and months of the year. In fact, if a dataset contains a string-based feature that is selected as input for a machine learning algorithm, those values are typically converted into integer-based values, which can be performed via the `map()` function in Pandas. Here are additional examples of qualitative data:

- audio (length)
- pictures or paintings (dimensions)
- text (word count/file size)
- video (length)

Since the items in the preceding list have a parenthetical term that can be used to "measure" the various items, why are they not considered quantifiable and therefore measurable, just like the earlier list? The main difference is that the items in the qualitative items are a form of multimedia, so they do not have a direct and immediate physical characteristic.

However, there are use cases in which media-related data *can* be treated as quantifiable. For example, suppose a company classifies ambient sounds. One practical scenario involves determining if a given sound is a gunshot versus the sound of a backfiring car. As such, the decibel level is an important quantifiable characteristic of both sounds.

In the case of paintings, it is certainly true that they can be "measured" by their selling price, which can sometimes be astronomical.

As another example, consider writers who are paid to write text-based documents. If their payment is based on the number of words in their documents, then the length of a document is a quantifiable characteristic. However, people who read articles typically do not make a distinction between an article that contains 400 words, 450 words, or 500 words.

Finally, the cost of generating a text document that contains the dialogue in a movie can be affected by the length of the movie, in which case videos have a quantifiable characteristic.

Types of Statistical Data

The preceding sections described several data types, whereas this section classifies data types from a statistical standpoint. There are four primary types of statistical data:

- nominal
- ordinal
- interval
- ratio

One way to remember these four types of statistical data is via the acronym NOIR (coincidentally the French word for "black"). Please refer to the earlier sections for details regarding any of these data types.

WORKING WITH DATA TYPES

If you have experience with programming languages, then you know that explicit data types exist (e.g., C, C++, Java, and TypeScript). Some programming languages, such as JavaScript and awk, do not require initializing variables with an explicit type: the type of a variable is inferred dynamically via an implicit type of system (i.e., one that is not directly exposed to a developer).

In machine learning, datasets can contain features that have different types of data, such as a combination of one or more of the following types of features:

- numeric data (integer/floating point and discrete/continuous)
- character/categorical data (different languages)
- date-related data (different formats)
- currency data (different formats)
- binary data (yes/no, 0/1, and so forth)
- nominal data (multiple unrelated values)
- ordinal data (multiple and related values)

Consider a dataset that contains real estate data, which can have 30 or more columns, often with the following features:

- number of bedrooms in a house: numeric value and a discrete value
- number of square feet: a numeric value and (probably) a continuous value
- name of the city: character data
- construction date: a date value
- selling price: a currency value and probably a continuous value
- "for sale" status: binary data (either "yes" or "no")

An example of nominal data is the seasons in a year: although many countries have four distinct seasons, some countries have two distinct seasons. However, seasons can be associated with different temperature ranges (summer versus winter). An example of ordinal data is an employee pay grade: 1=entry level, 2=one year of experience, and so forth. Another example of nominal data is a set of colors, such as {Red, Green, Blue}.

A familiar example of binary data is the pair {Male, Female}, and some datasets contain a feature with these two values. If such a feature is required for training a model, first convert {Male, Female} to a numeric counterpart (such as {0,1}), and a Pandas-based example is here:

```
df['gender'] = df['gender'].map({'Male': 0, 'Female':
1})
```

Similarly, if you need to include a feature whose values are the previous set of colors, you can replace {Red, Green, Blue} with the values {0,1,2}. Categorical data is discussed in more detail later in this chapter.

WHAT IS DRIFT?

In machine learning terms, *drift* refers to any type of change in distribution over a period of time. *Model drift* refers to a change (drift) in the accuracy of a model's prediction, whereas *data drift* refers to a change in the type of data that is collected. (Note that data drift is also called input drift, feature drift, or covariate drift.)

There are several factors that influence the value of data, such as accuracy, relevance, and age. For example, physical stores that sell mobile phones are much more likely to sell recent phone models than older models. In some cases, data drift occurs over a period of time, and in other cases, it is because some data is no longer relevant due to feature-related changes in an application. Always keep in mind that there might be multiple factors that can influence data drift in a specific dataset.

Two techniques for handling data drift are *domain classifier* and the *black-box shift detector*, both of which are discussed online:

https://blog.dataiku.com/towards-reliable-mlops-with-drift-detectors

In addition to the preceding types of drift, other types of changes can occur in a dataset, some of which are listed here:

- concept shift
- covariate shift
- domain shift
- prior probability shift
- spurious correlation shift
- subpopulation shift
- time shift
- For more information, please visit *https://arxiv.org/abs/1912.08142*.

Perform an online search to find more information about the topics in the preceding list of items. The following list contains links to open-source Python-based tools that provide drift detection:

- alibi-detect (*https://github.com/SeldonIO/alibi-detect*)
- evidently (*https://github.com/evidentlyai/evidently*)
- Torchdrift (*http://torchdrift.org/*)

DISCRETE DATA VERSUS CONTINUOUS DATA

In general, *discrete* data involves a set of values that can be counted whereas *continuous* data must be measured. Discrete data can reasonably fit in a drop-down list of values, but there is no exact value for making such a determination. One person might think that a list of 500 values is discrete, whereas another person might think it is continuous.

For example, the list of provinces of Canada and the list of states of the USA are discrete data values, but is the same true for the number of countries in the world (roughly 200) or for the number of languages in the world (more than 7,000)?

Values for temperature, humidity, and barometric pressure are considered *continuous* data types. Currency is also treated as continuous, even though there is a measurable difference between two consecutive values. The smallest unit of currency for US currency is one penny, which is 1/100th of a dollar (accounting-based measurements use the "mil," which is 1/1,000th of a dollar).

Continuous data types can have subtle differences. For example, someone who is 200 centimeters tall is twice as tall as someone who is 100 centimeters tall; similarly for 100 kilograms versus 50 kilograms. However, the temperature is different: 80 degrees Fahrenheit is not twice as hot as 40 degrees Fahrenheit.

Furthermore, the word "continuous" has a specific meaning in mathematics, and it is not necessarily the same as "continuous" in machine learning. In the former, a continuous variable (let's say in the 2D Euclidean plane) can have an uncountably infinite number of values. However, a feature in a dataset that can have more values that can be "reasonably" displayed in a drop-down list is treated *as though* it is a continuous variable.

For instance, values for stock prices are discrete: they must differ by at least a penny (or some other minimal unit of currency), which is to say, it is meaningless to say that the stock price changes by one-millionth of a penny. However, since there are "so many" possible stock values, it is treated as a continuous variable. The same comments apply to car mileage, ambient temperature, and barometric pressure.

BINNING DATA VALUES

The concept of *binning* refers to subdividing a set of values into multiple intervals, and then treating all the numbers in the same interval as though they had the same value. In addition, there are at least three techniques for binning data, as shown here:

- bins of equal widths
- bins of equal frequency
- bins based on k-means

(Source: *https://towardsdatascience.com/from-numerical-to-categorical-3252cf805ea2*)

As a simple example of bins of equal widths, suppose that a feature in a dataset contains the age of people in a dataset. The range of values is approximately between 0 and 120, and we could bin them into 12 equal intervals, where each consists of 10 values: 0 through 9, 10 through 19, 20 through 29, and so forth.

As another example, using quartiles is even more coarse-grained than the earlier age-related binning example. The issue with binning pertains to the unintended consequences of classifying people in different bins, even though they are closely related to each other. For instance, some people struggle financially because they earn a meager wage, but then they are also disqualified from financial assistance because their salary is higher than the cut-off point for receiving any assistance.

Scikit-learn provides the `KBinsKDiscretizer` class that uses a clustering algorithm for binning data:

https://scikit-learn.org/stable/modules/generated/scikit-learn.preprocessing.KBinsDiscretizer.html

A highly technical paper (PDF) with information about clustering and binning is online:

https://www.stat.cmu.edu/tr/tr870/tr870.pdf

Programmatic Binning Techniques

Earlier in this chapter, you saw a Pandas-based example of generating a histogram using data from the Titanic dataset. The number of bins was chosen on an ad hoc basis, with no relation to the data itself. However, there are several techniques that enable you to programmatically determine the optimal number of bins, some of which are shown here:

- Doane's Formula
- Freedman–Diaconis' Choice
- Rice's Rule
- Scott's Normal Reference Rule
- Square-Root Choice
- Sturge's Rule

Doane's Formula for calculating the number of bins depends on the number of observations n and the kurtosis of the data, and it is reproduced here:

```
k = 1 + log(n) + log(1 + kurtosis(data) * sqrt(n / 6.0))
```

Freedman–Diaconis' Choice specifies the number of bins for a sample x, and it is based on the InterQuartile Range (IQR) and the number of observations n, as shown in the following formula:

```
k = 2 * IQR(x)/[cube root of n]
```

Sturge's Rule to determine the number of bins k for Gaussian-based data is based on the number of observations n, and it is expressed as follows:

```
k = 1 + 3.322 * log n
```

In addition, after specifying the number of bins k, set the minimum bin width mbw as follows:

```
mbw = (Max Observed Value - Min Observed Value) / k
```

Experiment with the preceding formulas to determine which one provides the best visual display for your data. For more information about calculating the optimal number of bins, perform an online search for blog posts and articles.

Potential Issues When Binning Data Values

Partitioning the values of people's ages as described in the preceding section can be problematic. In particular, suppose that person A, person B, and person C are 29, 30, and 39 years old, respectively. Then, person A and person B are probably much more similar to each other than person B and person C, but because of the way in which the ages are partitioned, B is classified as closer to C than to A. In fact, binning can increase Type I errors (false positive) and Type II errors (false negative), as discussed in this blog post (along with some alternatives to binning):

https://medium.com/@peterflom/why-binning-continuous-data-is-almost-always-a-mistake-ad0b3a1d141f

CORRELATION

Correlation refers to the extent to which a pair of variables are related, which is a number between -1 and 1, inclusive. The most significant correlation values are -1, 0, and 1.

A correlation of 1 means that both variables increase and decrease in the same direction. A correlation of -1 means that both variables increase and decrease in the opposite direction. A correlation of 0 means that the variables are independent of each other.

Pandas has the `corr()` method, which generates a matrix containing the correlation between any pair of features in a data frame. Note that the diagonal values of this matrix are related to the variance of the features in the data frame.

A *correlation matrix* can be derived from a covariance matrix: each entry in the former matrix is a covariance value divided by the standard deviation of the two features in the row and column of a particular entry.

This concludes the portion of the chapter pertaining to dependencies among features in a dataset. The next section discusses different types of currencies that can appear in a dataset, along with a Python code sample for currency conversion.

What is a Good Correlation Value?

Although there is no exact value that determines whether a correlation is weak, moderate, or strong, there are some guidelines, as shown here:

- between 0.0 and 0.2: weak
- between 0.2 and 0.5: moderate
- between 0.5 and 0.7: moderately strong
- between 0.7 and 1.0: strong

The preceding ranges are for positive correlations, and the corresponding values for negative correlations are shown here:

- between -0.2 and 0: weak
- between -0.5 and -0.2: moderate
- between -0.7 and -0.5: moderately strong
- between -0.7 and -1.0: strong

However, treat the values in the preceding lists as guidelines. Some people classify values between 0.0 and 0.4 as weak correlations, and values between 0.8 and 1.0 as strong correlations. In addition, a correlation of 0.0 means that there is no correlation at all. (Or, perhaps, the correlation is considered extra weak?)

Discrimination Threshold

Logistic regression is based on the sigmoid function (which in turn involves Euler's constant) whereby any real number is mapped to a number in the interval (0,1). Consequently, logistic regression is well-suited for classifying binary class membership: i.e., data points that belong to one of two classes. For datasets that contain two class values, let's call them 0 and 1, logistic regression provides a probability that a data point belongs to class 1 or class 1, where the range of probability values includes all the numbers in the interval [0,1].

The *discrimination threshold* is the value whereby larger probabilities are associated with class 1 and smaller probabilities are associated with class 0. Some datasets have a discrimination threshold of 0.5, but in general, this value can be much closer to 0 or 1. Relevant examples include health-related datasets (healthy versus having cancer), sports events (win versus lose), and even the DMV's driving test, which requires 85% accuracy to pass in some US states.

Is a Zero Value Valid or Invalid?

In general, replacing a missing numeric value with zero is a risky choice: this value is obviously incorrect if the values of a feature are positive numbers between 1,000 and 5,000 (or some other range of positive numbers). For a feature that has numeric values, replacing a missing value with the mean of existing values can be better than the value zero (unless the average equals zero); also consider using the median value. For categorical data, consider using the mode to replace a missing value.

There are situations where you can use the mean of existing values to impute missing values but not the value zero, and vice versa. As an example, suppose that an attribute contains the height in centimeters of a set of persons. In this case, the mean could be a reasonable imputation, whereas 0 has the following problems:

1. It is an invalid value (nobody has height 0).

2. It will skew statistical quantities, such as the mean and variance.

You might be tempted to use the mean instead of 0 when the minimum allowable value is a positive number, but use caution when working with highly imbalanced datasets. As a second example, consider a small community of 50 residents with the following attributes:

1. 45 people have an average annual income of $50,000

2. 4 other residents have an annual income of $10,000,000

3. 1 resident has an unknown annual income

Although the preceding example might seem contrived, it is likely that the median income is preferable to the mean income, and certainly better than imputing a value of 0.

As a third example, suppose that a company generates weekly sales reports for multiple office branches, and a new office has been opened, but has yet to make any sales. In this case, the use of the mean to impute missing values for this branch would produce fictitious results. Hence, it makes sense to use the value 0 for all sales-related quantities, which will accurately reflect the sales-related status of the new branch.

WORKING WITH SYNTHETIC DATA

The ability to generate synthetic data (also called fake data) has practical uses, particularly in imbalanced datasets. Sometimes, it is necessary to generate synthetic data that closely approximates legitimate data because it is not possible to obtain actual data values.

For example, suppose that a dataset contains 1,000 rows of patient data in which 50 people have cancer and 950 people are healthy. This dataset is obviously imbalanced, and from a human standpoint, you *want* this dataset to be imbalanced (you want everyone to be healthy). Unfortunately, machine learning algorithms can be affected by imbalanced datasets, whereby they can "favor" the class that has more values (i.e., healthy individuals). There are several ways to mitigate the effect of imbalanced datasets, which are described in Chapter 2.

In the meantime, let's delve into the Python-based open-source library Faker for generating synthetic data, as discussed in the next section.

What is Faker?

The open-source `Python` library `Faker` is a very easy-to-use library that enables you to generate synthetic data, and it can be accessed at *https://pypi.org/ project/Faker/*.

On your machine, open a command shell and launch the following command:

```
pip3 install faker
```

After successfully installing `Faker`, you are ready to generate a dataset with synthetic data.

A Python Code Sample with Faker

Listing 1.1 displays the content of faker1.py that generates a synthetic name.

Listing 1.1: faker1.py

```
import faker

fake = faker.Faker()

name = fake.name()
print("fake name:",name)
```

Open a command shell and navigate to the directory that contains the file `faker1.py`. Launch the code with the following command:

```
python faker1.py
```

You will see the following output:

```
fake name: Dr. Laura Moore
```

Launching Faker from the Command Line

The previous section showed you a `Python` code sample for generating a synthetic name, and this section shows you how to generate synthetic values from the command line. Navigate to a command shell and type the following command to generate a synthetic name (lines that start with a "$" indicate commands for you to type):

```
$ faker address
96060 Hall Ridge Apt. 662
Brianton, IN 19597
```

```
$ faker address
8881 Amber Center Apt. 410
New Jameston, AZ 47448

$ faker name
Jessica Harvey
$ faker email
ray14@example.org

$ faker zipcode
45863

$ faker state
South Dakota

$ faker city
Sierrachester
```

As you can see, Faker generates different values for addresses, and does the same for other features (e.g., name and email). The next section shows you a systematic way to generate synthetic data and then save that data to a CSV file.

Generating and Saving Customer Data

Listing 1.2 displays the content of `gen_customers.py` that generates a set of customer names and saves them to a CSV file.

Listing 1.2: faker1.py

```
import os

# make sure we have an empty CSV file:
if os.path.exists(filename):
  os.remove(filename)
```

```python
else:
  print("File "+filename+" does not exist")

import pandas as pd
import faker

fake = faker.Faker()

# the name of the CSV file with customer data:
filename = "fake_customers.csv"

customer_ids = [100,200,300,400]

###############################################
# 1) loop through values in customer_ids
# 2) generate a customer record
# 3) append the record to the CSV file
###############################################

for cid in customer_ids:
  customer = [
      {
        "cust_id": cid,
        "name": fake.name(),
        "address": fake.street_address(),
        "email": fake.email()
      }
  ]

  # create a Pandas data frame with the customer
record:
  df = pd.DataFrame(data = customer )
```

```
  # append the generated customer record to the CSV
file:
  df.to_csv(filename, mode='a', index=False,
header=False)
```

Listing 1.2 starts by assigning a value to the variable `filename`, followed by a conditional block that checks whether the file already exists, in which case the file is removed from the file system. The next section contains several `import` statements, followed by initializing the variable `fake` as an instance of the `Faker` class.

The next section initializes the variable `customer_ids` with values for four customers, followed by a loop that iterates through the values in the `customer_ids`. During each iteration, the code creates a customer record that contains four attributes:

· a customer id (obtained from `customer_ids`)
· a synthetic name
· a synthetic street address
· a synthetic email address

The next portion of Listing 1.1 creates a `Pandas` data frame called `df`, which is initialized with the contents of the customer record, after which the data frame contents are appended to the `CSV` file specified near the beginning of Listing 1.1. Launch the code in Listing 1.1 by typing `python3 gen_customers.py` from the command line, and you will see the following type of output, which will be similar to (but not the same as) the output on your screen:

```
100,Jaime Peterson,17228 Kelli Cliffs Apt.
625,clinejohnathan@hotmail.com
200,Mark Smith,729 Cassandra Isle Apt. 768,brandon36@
hotmail.com
300,Patrick Pacheco,84460 Griffith Loaf,charles61@
proctor.com
400,Justin Owens,2221 Renee Villages,kyates@myers.com
```

Use the contents of Listing 1.2 as a template for your own data requirements, which involves changing the field types and the output `CSV` file.

The next section shows you a Python code sample that uses the `Faker` library to generate a `CSV` file called `fake_purch_orders.csv` containing synthetic purchase orders for each customer ID that is specified in Listing 1.2.

Generating Purchase Orders (Optional)

This section is marked "optional" because it is useful only if you need to generate synthetic data that is associated with data in another dataset. After customers register themselves in an application, they can have one or more associated purchase orders, where each purchase order is identified by the ID of the customer and an ID for the purchase order row.

Listing 1.3 displays the content of `gen_purch_orders.py` that shows you how to generate synthetic purchase orders for the list of customers in Listing 1.2 using the `Faker` library.

Listing 1.3: gen_purch_orders.py

```
filename = "fake_purch_orders.csv"

import os
if os.path.exists(filename):
  os.remove(filename)
else:
  print("File "+filename+" does not exist")

import pandas as pd
import numpy as np
import random
import faker

fake = faker.Faker()

##########################
# hard-coded values for:
# customers
# purchase orders ids
# purchased item  ids
##########################

customer_ids = [100,200,300,400]
purch_orders = [1100,1200,1300,1400]
```

```
item_ids     = [510,511,512]

outer_counter=1
outer_offset=0
outer_increment = 1000

inner_counter=1
inner_offset=0
inner_increment = 10

for cid in customer_ids:
  pid_outer_offset = outer_counter*outer_increment
  for pid in purch_orders:
    purch_inner_offset = pid_outer_offset+inner_
counter*inner_increment
    for item_id in item_ids:
      purch_order = [
          {
            "cust_id": cid,
            "purch_id": purch_inner_offset,
            "item_id": item_id,
          }
      ]
      df = pd.DataFrame(data = purch_order)
      df.to_csv(filename, mode='a', index=False,
header=False)
    inner_counter += 1
  outer_counter += 1
```

Listing 1.3 starts with code that is similar to Listing 1.2, followed by a code block that initializes the values for the variables customer_ids, purch_ orders, and item_ids that represent the ID values for customers, purchase orders, and purchased items, respectively. Keep in mind that these variables contain hard-coded values: in general, an application generates the values for customers and for their purchase orders.

The next portion of Listing 1.3 is a nested loop whose outer loop iterates through the values in the variable `customer_ids`, and for each ID, an inner loop iterates through the values in the variable `purch_orders`. Yet another nested loop iterates through the values in the `item_ids` variable.

The underlying assumption in this code sample is that every purchase order for every customer contains purchases for every item, which is usually not true. However, the purpose of this code sample is to generate synthetic data, which is not required to be identical to customer purchasing patterns. Fortunately, it *is* possible to modify the code in Listing 1.3 so that purchase orders contain a randomly selected subset of items, in case you need that level of randomness in the generated CSV file.

The remaining portion of Listing 1.3 works in the same manner as the corresponding code in Listing 1.2: each time a new purchase order is generated, a data frame is populated with the data in the purchase order, after which the contents of the data frame are appended to the CSV file that is specified near the beginning of Listing 1.3. Launch the code in Listing 1.3, and you will see the following type of output:

```
100,1010,510
100,1010,511
100,1010,512
100,1020,510
100,1020,511
100,1020,512
100,1030,510
100,1030,511
100,1030,512
100,1040,510
100,1040,511
100,1040,512
//details omitted for brevity
400,4130,510
400,4130,511
400,4130,512
400,4140,510
400,4140,511
400,4140,512
```

```
400,4150,510
400,4150,511
400,4150,512
400,4160,510
400,4160,511
400,4160,512
```

SUMMARY

This chapter started with a brief description of data literacy, as well as some aspects of EDA. Then you learned about common types of data, such a binary, nominal, ordinal, and categorical data. You also learned about continuous versus discrete data, quantitative and quantitative data, and types of statistical data.

In addition, you learned about concepts such as data drift and data leakage. Next, you learned about processing categorical data, and how to map categorical data to numeric data. Finally, you learned how to use the Python-based library called Faker to generate datasets containing synthetic data.

INTRODUCTION TO PROBABILITY

This chapter introduces you to some concepts in probability, recursion, and combinatorics. The first half of this chapter is about probability, followed by one-quarter of the chapter that involves recursion, and a final short section about combinatorics.

The first section of this chapter starts with introductory material regarding set theory, open and closed sets, and Brouer's theorem. Although this section is not vitally important to the remainder of this chapter, it is worth skimming through to prepare yourself for online articles involve these topics.

As always, you can skip the sections that are already familiar to you, but they may be worthwhile reading, as there may be some information that is new.

WHAT IS SET THEORY?

A *set* is a collection of objects. *Set theory* is the study of sets, which is relevant to probability because it is the measure of the size of a set. As a starting point, suppose that the set S is a collection of elements {s1, s2, . . ., sn} that can be finite or infinite.

Set B is a subset of A if every element in B is also an element of A. If B != A, then B is called a *proper* subset of A. An *empty set* is a set that does not contain any elements, often denoted as {}. An empty set is not the same as the set {0}, which has one element called "0." The empty set {} has the cardinality of 0, whereas the set {0} has the cardinality of 1.

Properties of Sets

Suppose that we have the sets A, B, and C, as well as the set Ac, which the complement of the set A. Then the following formulas are true:

- P(B∩Ac)=P(B)−P(A∩B)
- P(A∪B)=P(A)+P(B)−P(A∩B)
- If A⊂B, then P(A)≤P(B)

The *union* of two sets A and B is {x|x is in A *or* x is in B}, whereas the *intersection* of two sets A and B is {x|x is in A *and* x is in B}. The sets A and B are *disjoint* if they have no elements in common, so A intersect B = {}. A *partition* of the set A is a collection of sets {A1, A2, . . . , An} such that

1. Ai and Aj are pairwise disjoint
2. the union of {A1, A2, . . . , An} equals A

The *complement* of set A is the set of elements in a superset of A that is not in A itself. The difference A−B of two sets A and B is {x|x is in A *and* x is *not* in B}. Finally, two sets A and B are *independent* if their intersection is empty.

DeMorgan's Laws

DeMorgan's laws are the counterpart of the distributed laws for addition and multiplication that are shown here:

- (x−y)*z = x*z − y*z
- (x+y)*z = x*z + y*z

Treat union and intersection as the counterparts to subtraction and addition in the following formulas:

- (A ∩ B)C = AC ∪ BC
- (A ∪ B)C = AC ∩ BC

Sets, Sample Spaces, and Events

A set S is called a *sample space* if S is exhaustive and exclusive. Specifically, a sample space must contain all possible outcomes and no duplicate elements. For example, if you toss a six-sided die that has labels S = {1,2,3,4,5,6} for its six sides, then S is a sample space because it contains all possible outcomes from tossing a die. By contrast, the set {1,2,3,4,5} is not a sample space because it excludes one of the possible outcomes. Finally, an *event* is a subset of a sample space.

OPEN, CLOSED, COMPACT, AND CONVEX SETS (OPTIONAL)

This section is purely optional and can be skipped with no loss of continuity in this chapter. However, this material is relevant when you want to learn more advanced concepts in probability.

With the previous points in mind, the interval `[a,b]` where a `<=` b is a *closed set*. The union of `[a,b]` and `[c,d]` where a `<=` b and b `<` c and c `<=d` is also a closed set (i.e., it is the union of two closed sets).

A set `S` that is a subset of the real number line is a *convex set* if the following is true: given any pair of elements `x` and `y` that belong to `S`, then all the points on the line that connect `x` and `y` are also in `S`. For example, `[a,b]` with a<=b is convex, whereas the set `S` that is the union of `[a,b]` and `[c,d]`, where b `<` c, is not convex: the point `(b+c)/2` lies outside the set `S`.

The set `(a,b)` where a<b is an *open set*. The union of `(a,b)` and `(c,d)` where a<b and b `<` c and c<d is also an open set (i.e., it is the union of two open sets). However, the union of `(a,b)` and `(c,d)` where a<b and b `<` c and c<d is *not* convex because the point `(b+c)/2` lies outside the set `S`.

Now suppose that the set `U` consists of a collection of open sets, where each of those open sets is a subset of a set `X`. Then `U` is an *open covering* of a set `S` if any element `X` in `S` belongs to at least one element of `U`.

A closed set `S` is *compact* if every open covering `U` of `S` contains a finite subcovering. In slightly different words, this means that any covering `U` that contains an infinite number of elements can be replaced by a finite number of elements of `U` that *still* form an open covering of the set `S`.

For example, the set `[a,b]` is compact, and the set consisting of the union of two closed sets `[a,b]` and `[c,d]` where a `<` b and b `<` c is also a compact set. However, the set `[a,b)` is *not* compact, and neither is the set consisting of the union of `[a,b)` and `(c,d]`, where a `<` b and b `<` c. In fact, the latter pair of sets are neither open nor closed.

Now suppose that `x1` and `x2` are real numbers. A *linear combination* of `x1` and `x2` is expressed as follows:

```
a1*x1 + (1-a1)*x2 where 0 <= a1 <= 1.
```

Now suppose that `S` is a subset of the real number line with the following property: given any `x1`, `x2` in `S`, if `a1*x1 + (1-a1)*x2` belongs to `S` for any value of `a1` that satisfies `0 <= a1 <= 1`, then the set `S` is called a *convex set*. For example, `S1 = [0,1]` is a convex set whereas `S2 = [0,1] U [2,3]` is not a convex set.

Let's continue with the notion of convexity in the case of real-valued functions in the Euclidean plane. Suppose that the function `f:X -> R` is a function that maps the convex subset `X` (of a vector space) to the real number line. Then `f` is a convex function if the following is true:

```
f(t*x1 + (1-t)*x2) <= t*f(x1) + (1-t)*f(x2) for any t
in 0 <= t <= 1.
```

For example, `f(x) = x*x` for any real number `x` is a convex function, whereas `f(x) = 4 - x*x` is not a convex function.

The Law of Large Numbers

The Law of Large Numbers states that as the size of a sample set from a population increases, the higher the accuracy of an estimate the sample mean will be of the population mean.

CONCEPTS IN PROBABILITY

The theoretical probability for an event `A` can be calculated as follows:

```
P(A) = (Number of outcomes favorable to Event A) /
(Number of all possible outcome)
```

If you have ever performed a science experiment in one of your classes, you might remember that measurements have some uncertainty. In general, we assume that there is a correct value, and we endeavor to find the best estimate of that value. Before we delve into examples of probabilities, let's briefly discuss some terminology, which is the topic of the next section.

Basic Terminology

There are some fundamental concepts in probability theory, some of which are listed here:

- The *frequency* is a non-negative number that specifies how often an outcome occurs.
- A *distribution* is a mapping from each outcome to the frequency of that outcome.

- A *probability distribution* is a distribution of positive numbers (typically fractions or decimal values) whose sum equals 1.

For example, the distributions {1/2, 1/2} and {1/6, 1/6, 1/6, 1/6, 1/6, 1/6} are both discrete probability distributions because the values in each distribution are greater than 0 and the sum of the values equals 1. Probability distributions are used in calculating the expected value of random variables, which is discussed later in the chapter.

When we work with an event that can have multiple outcomes, we define the probability of an outcome as the chance that it will occur, which is calculated as follows:

```
p(outcome)  = (# of times outcome occurs)/(total number
of outcomes)
```

For example, in the case of a single balanced coin, the probability of tossing heads H equals the probability of tossing tails T:

```
p(H) = 1/2 = p(T)
```

The set of probabilities associated with the outcomes {H, T} is shown in the set P:

```
P = {1/2, 1/2}
```

One other concept is called *conditional probability*, which refers to the likelihood of the occurrence of event E1 given that event E2 has occurred. A simple example is the following statement:

```
"If it rains (E2), then I will carry an umbrella (E1)."
```

Four Types of Item Selection

Later in this chapter, we will discuss tasks such as "selecting a ball from a jar," which can be performed *with* replacement (i.e., you replace the ball in the jar) or *without* replacement (you do not return the ball to the jar).

In addition, it is possible to distinguish between the items in a given selection sequence, or you can ignore the selection sequence. For example, if a selection sequence of colored balls consists of R-G-B, you can treat this sequence to be the same as the sequence of colors B-R-G, or you can consider them to be identical (it is up to you or the task that you are given). In summary, here are the four possible selection types:

- Select *with* replacement and the ordering *is* significant
- Select *with* replacement and the ordering *is not* significant

• Select *without* replacement and the ordering *is* significant
• Select *without* replacement and the ordering *is not* significant

For more information, please see the following URL:

https://towardsdatascience.com/key-concepts-to-improve-your-under-standing-of-probability-theory-ca1d999dd6c9

Unless stated otherwise, the selection tasks in this chapter do not consider the ordering to be significant.

Probability of Item Selection (Binary Case)

Some experiments involve selecting items with replacement, which differs when items are selected without replacement. For example, card games involve selecting cards without replacement, whereas selecting items from a bucket can be performed with replacement as well as without replacement.

For example, suppose that a bucket contains 10 red balls and 10 green balls. What is the probability that a randomly selected ball is red? The answer is `10/(10+10) = 1/2`. What is the probability that the second ball is also red?

There are two scenarios with two different answers (probabilities). If each ball is selected *with* replacement, then each ball is returned to the urn after selection, which means that the urn always contains 10 red balls and 10 green balls. In this case, the answer is `1/2 * 1/2 = 1/4`. In fact, the probability of any event is independent of all previous events.

However, if balls are selected without replacement, then the answer is `10/20 * 9/19`. As you undoubtedly know, card games are also examples of selecting cards without replacement.

Probability of Item Selection (Multi-Class Case)

Suppose that a bucket contains 5 red balls, 3 green balls, and 2 white balls. The probabilities of selecting certain balls (with replacement) from the bucket are as follows:

```
p(red)   = 5/10
p(green) = 3/10
p(white) = 2/10
```

Based on an earlier section, the set of numbers S = {5/10, 3/10, 2/10} is a probability distribution, which can be used to calculate the associated entropy and Gini impurity. The entropy H and Gini impurity are calculated as follows:

```
     3
H = SUM pi * log(pi)
    I=1
```

```
       3                      3
Gini = SUM pi * (1- pi) = 1 - SUM pi**2
      i=1                    i=1
```

For the general case, replace 3 with the variable n, which represents the number of probabilities for the number of items involved.

Joint Probability

Given events X and Y, the joint probability of X and Y is the probability that both occur in the same trial or experiment. For example, suppose we toss a fair coin twice, and we want to calculate the following probabilities:

- P(Heads and Heads)
- P(Heads and Tails)
- P(Tails and Tails)
- P(Tails and Heads)

Since the coin tosses are independent of each other, and P(Heads) = P(Tails) = 1/2, the answer to all four cases is 1/2 * 1/2 = 1/4.

As another example, what is the probability of selecting a five of diamonds from a deck of cards? In this case, let X = P(selecting a 5) and let Y = P(selecting a diamond). Consequently, we have P(X) = 4/52 and P(Y) = 13/52, so P(X and Y) = 4/52 * 13/52 = 52/[52*52] = 1/52.

Another (simpler) calculation involves the observation that there is only one five of diamonds in a deck of 52 distinct cards, so the probability must be 1/52.

Mutually Exclusive Events

Suppose you know that the elements in the set S = {S1, S2, S3, S4, S5} have the following probabilities of occurrence:

```
p(S1) = 0.10
p(S2) = 0.10
```

```
p(S3) = 0.15
p(S4) = 0.25
p(S5) = 0.40
```

If we assume that the events in set S are mutually exclusive, then the probability of event S1 or S2 occurring equals `0.10 + 0.10 = 0.20`. The probability of S4 and S5 *not* occurring equals `1 - (0.25 + 0.40) = 0.35`. In general, let S = {S1, S2, ..., Sn} be a set of mutually independent events. Then, the probability of the union of the events in S is calculated as follows:

```
                        n
P(S1 u S2 u ... u Sn) = SUM P(Si)
                       I=1
```

Probability, Likelihood, and Odds

Suppose that the set S consists of a finite set of outcomes, and A is a subset of S. Then the probability P(A) of event A is defined as follows, where |A| and |S| are the cardinality of the sets A and S:

```
P(A) = |A|/|S|
```

If a hypothesis H (a topic in Chapter 5) is also provided, then the probability of event A is defined as follows:

```
P(A|H) = P(A intersect H)/P(H)
```

The *odds* of an event A are defined as the ratio of |A| and |S|-|A|. Informally, the odds equal the number of ways that an event A *can occur* compared to the number of ways that event A *does not* occur. For example, if the odds of winning a match are 3:2, then there are 3 out of 5 possibilities of winning a match, which equals a probability of `3/(3+2) = 3/5 = 60%`.

SET THEORY AND PROBABILITY

In simplified terms, probability involves the calculation of the ratio of the cardinality of sets. The set of possible outcomes S is called the *outcome space*, and an event is a subset E of the set S.

As a simple illustration of a set-oriented perspective for calculating probabilities, suppose that you toss a fair coin three times, and you want to determine the probability of tossing exactly one heads. There are 8 possible outcomes, as shown here:

```
HHH
HHT
HTH
HTT
THH
THT
TTH
TTT
```

The set of all possible sequences has a cardinality of 8, and the set of sequences with exactly one heads (shown in bold in the preceding list) has a cardinality of 3. Since all sequences have equal probability of occurrence, the desired probability is 3/8. Obviously, the preceding example is very simple and intuitive, yet it is the basis for calculating probabilities that involve multiple subsets, such as one of the ball-selection tasks that you will see later in this chapter.

In general, similar logic is used in the next several sections that contain probability-related questions in these areas:

- coin tossing
- dice tossing
- card selecting
- container/ball selection
- gender-related probabilities

Of course, the underlying logic for the preceding types of tasks can also be adapted to other tasks that involve selection, either with replacement or without replacement. Now let's consider some coin tossing problems, which is the topic of the next section.

COIN TOSSING PROBABILITIES

Earlier you learned that for a single balanced coin, the probability of tossing heads H equals the probability of tossing tails T:

```
p(H) = 1/2 = p(T)
```

If you toss a balanced coin twice, there are four possible outcomes, as shown here:

- `HH`
- `HT`
- `TH`
- `TT`

Each of the four preceding outcomes can occur with probability `1/2*1/2 = 1/4` because coin tosses are independent of each other. Similarly, the outcomes of tossing a coin three times are shown here:

- `HHH`
- `HHT`
- `HTH`
- `HTT`
- `THH`
- `THT`
- `TTH`
- `TTT`

Each of the eight preceding outcomes can occur with probability `1/2*1/2*1/2 = 1/8` because coin tosses are independent of each other. In general, if you toss a coin n times, there are `2**n` possible sequences, and each one can occur with probability `1/2**n`.

With the preceding points in mind, what is the probability of tossing a coin ten times and a heads appearing only once? Here are the ten possible sequences that contain one heads:

```
HTTTTTTTTT
THTTTTTTTT
TTHTTTTTTT

...

TTTTTTTTTH
```

Therefore, the answer is `10/2^10 = 10/1024`, because each sequence has the same probability of occurrence (i.e., `1/1024`).

Notice that `10 = C(10,1)`, where `C(n,k)` is the number of ways of selecting k objects from a set of n objects. Consequently, the probability p of tossing a coin ten times and heads appearing twice equals the following:

```
p = (# of sequences with two heads)/(total # of
sequences)
```

```
= C(10,2) / 1024
= (10*9/2)/ 1024
= 45 /1024
```

Finally, we can use an earlier observation to conclude that the probability of tossing a coin n times and a heads appearing k times equals C(n,k)/2^n.

Threshold Probabilities

How many coin tosses are required to toss heads with 80% probability? Clearly, you need to toss a coin more than once, because one coin toss gives you only 50% probability. If you toss a coin twice, the probability of tossing heads is computed as follows:

```
P(Heads on second toss) =
P(Tails on first toss) * P(Heads on second toss) = 1/2
* 1/2 = 1/4
```

Hence, the probability of tossing heads in the first two coin tosses = 1/2+1/4 = 3/4.

Extending the preceding logic that involves tossing heads during exactly two coin tosses, the probability of tossing heads in three coin tosses is calculated as follows:

```
P(Heads on third toss) = P(Tails) * P(Tails) * P(Heads)
= 1/2*1/2*1/2 = 1/8
```

Thus, tossing heads during three coin tosses = 3/4+1/8 = 7/8 = 0.875, which answers the question posed above.

Notice that the sequence of probabilities is the cumulative sum of a partial geometric series. If we toss a coin n times then the probability p of tossing heads is here:

```
p = 1/2 + 1/4 + 1/8 + . . . + 1/2^n
  = 1/2*[1+ 1/2 + . . . + 1/2^(n-1)]
  = 1/2*[(1-(1/2)^n)/(1/2)]
  = 1-(1/2)^n
```

Now we can solve the preceding equation for n as follows:
```
p = 1-(1/2)^n
(1/2)^n = 1-p
n*(-1) = log(1-p)
n = (-1)*log(1-p)
```

Note that the logarithm in the preceding equation is base 2, and that n is an integer. We set n equal to the value `ceil[(-1)*log(1-p)]`. As an example, if we specify p = 80% as we did in the previous example, then we have the following:

```
n = (-1)*log(1-p)
  = (-1)log(0.2)
  = (-1)*log(2/10)
  = (-1)*[log 2 - log 10]
  = log 10 - log 2
  = 3.322 - 1 = 2.322
```

Hence, n equals the next integer larger than 2.322 (the "ceiling"), which equals 3, and this latter value confirms what we calculated manually.

The advantage of the formula is that we can calculate the value of n for any probability, some of which would be impractical to compute in a manual fashion. For example, if we want a 99.99% probability of tossing heads, the value for n is calculated as follows:

```
n = (-1)*log(1-p)
  = (-1)*log(1-0.999)
  = (-1)*log(0.001)
  = (-1)*log(1/1000)
  = (-1)*[log 1 - log 1000]
  = [log 1000 - log 1]
  = 9.966 - 0 = 9.966
```

Hence, n equals the next integer larger than 9.966, which equals 10. Did you expect the answer to be significantly larger?

Now consider the following coin-tossing question: what is the probability p of tossing a coin ten times and *at most* 9 heads appear? First, calculate the probability that a sequence of 10 coin tosses contains a *single* tails, which is the same as the probability that a sequence of 10 coin tosses contains a single heads. We calculated the latter in a previous section, and the probability is `10/2^10`. Thus, the probability p of tossing a coin ten times and *at most* 9 heads appear equals `1 - 10/2^10 = 1014/1024`.

If we want the probability p of tossing a coin ten times and *at most* 8 heads appear, we can subtract the number of occurrences of exactly one heads and the number of occurrences of exactly two heads from the value 1, as shown here:

```
1 - [C(10,1) + C(10,2]/2^10 = 1 - (10+45)/1024  =
964/1024
```

Yet another question: what is the probability p of tossing a coin ten times and *an even number* of heads appear? The answer is a sum of combinatorial coefficients, as shown here:

```
p = [C(10,0) + C(10,2) + . . . + C(10,10)]/1024
```

If we want to determine the probability p of tossing a coin ten times and *an odd number* of heads appears, the answer is a sum of combinatorial coefficients, as shown here:

```
p = [C(10,1) + C(10,3) + . . . + C(10,9)]/1024
```

Incidentally, this chapter contains the derivation for the sum of the first n terms of an arithmetic expression, as well as the first n terms of a geometric expression.

Solving "At Least" and "At Most" Tasks

Suppose you toss a well-balanced coin three times, and you want to know the probability that you will toss at least one heads. Based on the comments in previous sections, the probability of tossing a heads or a tails are both equal to `1/2 (= 0.5)`, and that the outcome of coin tosses are independent of each other. Therefore, tossing the sequence `TTT` equals `(1/2)*(1/2)*(1/2) = 1/8`, which is the probability of tossing any of the following sequences:

```
HHH
HHT
HTH
HTT
THH
THT
TTH
TTT
```

The sum of tossing any of the preceding sequences equals `8 * (1/8) = 1`, and tossing at least one heads in three consecutive coin tosses only excludes the sequence `TTT`, which occurs with probability `1/8`. Therefore, the probability of tossing at least one heads is `1 - 1/8 = 7/8`.

Similarly, tossing at most two heads excludes the sequence HHH, which occurs with probability 1/8 (the logic is similar to the sequence TTT), and therefore the answer is the same as the previous case: 1 - 1/8 = 7/8.

In general, we can replace three coin tosses with n coin tosses, which means that we have 1/2^n instead of 1/8 in both the previous tasks. Consequently, the probability of tossing at least one heads in n coin tosses equals the probability of tossing at most (n-1) heads in n coin tosses equals, which equals 1 - 1/2^n = (2^n - 1)/2^n.

Solving "At Least 2" and "At Most (n-2)" Tasks

Suppose you toss a well-balanced coin n times, and you want to know the probability that you will toss at least two heads. First, the number distinct sequences of heads and tails that can occur in sequences of length n equals 2*2* . . . *2 = 2^n. Next, we need to calculate the probability of tossing zero heads and the probability of tossing one heads, compute the sum of those two probabilities, and then subtract that sum from 1.

We already know that tossing zero heads is the probability of tossing the sequence of length n consisting entirely of the value T, which equals (1/2)*(1/2)* . . . *(1/2) = 1/2^n. Next, the probability of tossing exactly one heads is the probability of tossing any sequence of length n that contains exactly one H. There are n such sequences: each of these sequences is represented by a row in the nxn identity matrix. The probability of n such sequences is n/2^n.

Hence, the probability of tossing at most (n-2) heads using a well-balanced coin equals

```
1 - (1/2^n + n/2^n) = 1 - (n+1)/2^n = (2^n - n -
1)/2^n.
```

If you are unfamiliar with (or have forgotten) binomial coefficients, you can read online articles that explain why there are C(n,2) ways to select two objects from a set of n distinct objects.

Consequently, there are 1, n, and C(n,2) ways to toss zero heads, one heads, and two heads, respectively, in a sequence of n coin tosses using a well-balanced coin. Hence, the number of ways to toss at least three heads (n is at least 3) is calculated as follows:

```
1 - (1 + n + C(n,2))/2^n = 1 - [1+n+n*(n-1)/2]/2^n
```

Furthermore, the preceding expression is the probability of tossing *at most* (n-3) heads in in a sequence of n coin tosses using a well-balanced coin. The logic for this result is similar to the logic of the corresponding case in the preceding section.

One other observation is that the calculated terms can be expressed as binomial coefficients:

```
1 + n + C(n,2) = C(n,0) + C(n,1) + C(n,2)
1 + n + C(n,2) = C(n,n) + C(n,n-1) + C(n,n-2)
```

In fact, for any non-negative integers n and k (and k <= n) the following result is true:

```
C(n,k) = C(n,n-k)
```

The examples in this section are easy to solve because there are only a few cases to exclude, all of which have probabilities that are easily calculated. By contrast, suppose that you want to calculate the probability that at most 9 heads appear in a sequence of 10 coin tosses. The latter task is less straightforward, but there is a way to define an expression for the answer.

Based on the concepts in the previous examples, the answer involves excluding the sequences that contain 0, 1, 2, . . . or 9 heads, whose sum we can express as follows:

```
           9
S(9)   = SUM C(9,k)
          k=0
```

Now, divide the preceding sum by 2^10 because there are 2^10 distinct sequences of length 10 that can occur, and then and subtract from 1, which equals 1 - S(9)/2^10. This result can be generalized by replacing 9 with k as follows:

```
          n-1
S(n-1) = SUM C(n,k)
          k=0
```

DICE TOSSING PROBABILITIES

Given a well-balanced die, the probability of tossing any value in the set {1,2,3,4,5,6} is 1/6. In addition, the probability of the individual outcomes in a sequence of values are all independent of each other, which is analogous to the result of a sequence of coin tosses for a fair coin.

Solving "At Least" and "At Most" Tasks

Suppose you toss two well-balanced dice, and you want to know the probability that you will toss at least one 6. As you learned in the previous section, the probability of tossing a 6 for a single die equals 1/6, and therefore the probability of not tossing the value 6 equals 5/6.

Hence, the probability of tossing two dice with no occurrence of a 6 is the product of 5/6 with 5/6, each of which is the probability of a "non-six." Therefore, the answer equals `5/6*5/6 = 25/36`.

However, the probability of tossing two dice with a *single* occurrence of a 6 is `5/36 + 5/36 = 10/36` because there are 10 pairs of values that contain a single 6, as shown here:

```
(6,1)
(6,2)
(6,3)
(6,4)
(6,5)
(1,6)
(2,6)
(3,6)
(4,6)
(5,6)
```

Thus, the probability of getting at least one 6 from tossing a pair of dice equals `1 - 25/36 = 10/36`. Note that the pair `(6,6)` is excluded because it contains *two* sixes. Another variation involves calculating the probability of tossing exactly two sixes, which equals `1/6 * 1/6 = 1/36`.

Yet another variation involves finding the probability of tossing at least one six: the answer equals the probability of tossing *at most one* six plus the probability of tossing *exactly two* sixes. Hence, the answer is the sum of `10/36` and `1/36` which equals `11/36`.

Solving "At Least 2" and "At Most (n-2)" Tasks

Suppose you toss two well-balanced dice, and you want to know the probability that you will toss at least two occurrences of 6. As you learned in the previous section, the probability of tossing a 6 for a single die equals 1/6, and therefore the probability of not tossing the value 6 equals 5/6.

Similarly, the probability of tossing a pair of dice and getting at least two occurrences of 6 equals `1 - (zero occurrences of 6 plus 1 occurrence of 6) = 1 - (25/36 + 10/36) = 1 - 35/36 = 1/36`. Another way to derive the same result is even simpler: the only way to get two occurrences of 6 from tossing a pair of dice is the pair `(6,6)`, which has a probability of 1/36.

Find the probability of rolling an even number when you roll a die containing the numbers `1-6`. The answer is straightforward: there are 6 possible outcomes, including three even values and three odd values, all of which occur with equal probability `1/6`, so `the probability of tossing an even number = 3/6 = the probability of tossing an odd number.`

Solving Unequal Dice Values

The following tasks are useful for solving "complementary" tasks: finding unequal pairs is the complement of finding equal pairs, and the latter is often very simple.

Task #1: Suppose you toss two well-balanced dice. Calculate the probability that the dice have a) the same values and the probability that the dice have b) different values.

First, there are `6x6 = 36` possible pairs of values, and there are six pairs that are equal, as shown here:

```
(1,1)
(2,2)
(3,3)
(4,4)
(5,5)
(6,6)
```

Hence, the probability of equal values in a pair is `6/36`. Moreover, there are `36 - 6` remaining pairs whose values are different in each pair, and therefore the probability of different values in a pair equals `30/36`.

Task #2: Suppose you toss two well-balanced dice. Calculate the probability that the first die has larger values than the second die.

In Task #1, the unequal pairs are symmetric, so if `(m,n)` is an unequal pair, then `(n,m)` is also an unequal pair. Therefore, there is an equal number of pairs in which `m > n` as there are pairs in which `m < n`. Thus, the answer is `15/36`.

Task #3: Suppose you toss three well-balanced dice. Calculate the probability that they have different values. First, notice that there are `6x6x6 = 216` possible triples.

Next, there are two possible cases. The first case only requires that no triple contains equal values, which means that `(1,1,1)` is excluded but `(1,1,3)` is considered unequal. Notice that there are only six triples in which all the values are the same, as shown here:

`(1,1,1)`

`(2,2,2)`

`(3,3,3)` .

`(4,4,4)`

`(5,5,5)`

`(6,6,6)`

If you roll three dice, then the number of triples equals `6*6*6 = 216`, so the probability that all values are equal is `6/216`, whereas the probability that at least one value in a triple is different equals `(216-6)/216 = 210/216`.

In the second case, the values in a triple must be pairwise distinct, which means that `(1,2,3)` is valid whereas `(1,1,3)` is not a distinct triple. To find the solution, let's suppose that the value of the first die is 6, which means that the second die and third die have values in the range of 1 through 5 inclusive, *and* must also be pairwise distinct. In a square whose side has length 5, there are 5 integer-valued pairs that are equal, which means that there are 25-5 pairs that are distinct.

Repeat the same process for the value of the first die equal to 5, 4, 3, 2, and 1. In the case of 5, the other two dice can have any value between 1 and 6 *except* 5, which means that there are 25 possible pairs and 5 pairs are equal. Once again, there are 20 distinct pairs, and they are pairwise different from the initial value of 5.

Since there are 6 values for the first die, and 20 distinct values for the second and third dice, there are `6x20 = 120` unique triples, so the probability of throwing three dice whose values are pairwise distinct equals `120/216`.

Working with k-Sided Dice

Let's find the number of distinct k-tuples of values that are possible when you toss n balanced dice with k sides, where distinct values means that at least one pair in the k-tuple is different. This task corresponds to the previous task that involved subtracting the number of k-tuples with all values equal from the total number of k-tuples. So, we now have `k^n - k` instead of `6^2 - 6`, and since there are `k^n` possible tuples, the answer is `(k^n - k)/k^n`.

Now, find the number of pairwise distinct k-tuples of values that are possible when you toss n balanced dice with k sides. This task corresponds to the previous task that involved subtracting the number of k-tuples that have equal pairwise values. Replace `6` by `k` and `5` by `(k-1)` to generalize the previous scenario, and the answer is as follows:

`(k*[(k-1)*(k-1)-(k-1)])/k^n = (k*[(k-1)*(k-2)])/k^n`

CARD DRAWING PROBABILITIES

Consider the experiment of drawing two cards without replacement from a well-shuffled deck of 52 cards. What is the probability that the pair of cards contains the following?

- no aces (S1)
- exactly one ace (S2)
- exactly two aces (S3)
- at least one ace (S4)

Scenario S1 (no aces):

There are 4 aces in a standard deck of 52 cards, which means that the probability of drawing a non-ace for the *first* card is 48/52 (i.e., one of the 48 non-ace cards).

Next, the probability of drawing a non-ace for the *second* card is 47/51 (i.e., one of the 47 non-ace cards). Hence, the probability of drawing zero aces without replacement is 48/52 ° 47/51.

Scenario S2 (exactly one ace):

There are two possible sequences: AN and NA, where A refers to an ace and N refers to a non-ace card. The probability of drawing an ace for the first card is 4/52, and the probability of drawing a non-ace for the second card is 47/51 (i.e., one of the 47 non-ace cards). Hence, the probability of drawing AN without replacement is `4/52 * 47/51`.

For the NA sequence, the probability of drawing a non-ace for the first card is 48/52, and the probability of drawing an ace for the second card is 4/51 (i.e., one of the 4 ace cards). Hence, the probability of drawing NA without replacement is `48/52 * 4/51`. The final answer is the sum of the two prob-abilities: `4/52*47/51 + 48/52*4/51`.

Scenario S3 (exactly two aces):

There is only one possible sequence: `AA`, where `A` refers to an ace. The probability of drawing an ace for the first card is 4/52, and the probability of drawing an ace for the second card is 3/51 (i.e., one of the 3 remaining ace cards). Hence, the probability of drawing `AA` without replacement is `4/52 * 3/51`.

Scenario S3 (at least one ace):

There are the following possible sequences: `AN`, `NA`, and `NN`. As explained earlier, we have the following probabilities:

- an `AN` sequence has the probability `4/52 * 47/51`
- an `NA` sequence has the probability `4/52 * 47/51 + 48/52*4/51`
- an `AA` sequence has the probability `4/52 * 3/51`

The final answer involves the sum of the preceding probabilities, which can be calculated as follows:

```
4/52*[47/51 + 47/51 + 3/51 + 12/52*4/51]
= 4/52*[97/51 + 48/51]
= 4/51*[145/52]
= 0.2186
```

Draw Cards of the Same Suit

Suppose you draw a card from a deck of 52 cards. Since there are 12 other cards in the deck of 51 cards that have the same suit, the probability of draw-ing a second card of the same suit is 12/51.

If you extend this problem to the probability of drawing three cards of the same suit, then the probability of drawing the third card with the same suit as the first two cards is 11/50. Therefore, drawing three cards of the same suit has a probability of `12/51*11/50`.

Draw No-Pair Cards

The first card from a card deck is drawn arbitrarily with a probability of 1. The second card cannot match the first card, which means that the second card must have a different numeric value or a different face value.

For example, if the first card is the five of diamonds, then the other three cards with the value five must be avoided when selecting the second card. Hence, the second card can be chosen from (51-3) cards. Therefore, the probability of drawing two no-pair cards equals `1 * 48/51 = 48/51`.

Draw Equal Rank Cards

This task is the "opposite" of the task in the previous section. Specifically, suppose you draw a card from a deck of 52 cards. Since there are 3 other cards in the deck of 51 cards that have the same rank, the probability of drawing a second card of the same rank is 3/51.

If you extend this problem to the probability of drawing three cards of the same rank, then the probability of drawing the third card with the same rank as the first two cards is 2/50. Drawing three cards of the same rank has a probability of `3/51*2/50`. Finally, the probability of drawing four cards of the same rank is `3/51*2/50*1/49`.

You can extend this task in interesting ways. For example, try to calculate the probability of drawing the following:

- three of a kind and a pair (e.g., three sevens)
- two pairs (e.g., two fives and two nines)
- three of a kind and a pair

CONTAINER-BASED PROBABILITIES

Suppose that a jar contains 5 blue marbles, 3 red marbles, and 2 green marbles. There are 10 possible outcomes because the total number of marbles equals 10. Clearly the probability of selecting a blue marble, or selecting a red marble, or selecting a green marble is 5/10, 3/10, and 2/10, respectively.

The probability of randomly selecting anything other than a blue marble is 5/10. The probability of selecting a non-red or a non-green marble from the jar is 7/10 and 8/10, respectively.

The probability of randomly selecting two consecutive blue marbles is `5/10 * 4/9`. The probability of randomly selecting three consecutive blue marbles is `5/10 * 4/9 * 3/8`.

The probability of randomly selecting a blue marble followed by a red marble followed by a green marble is `5/10 * 3/9 * 2/8`.

Now consider the probability of randomly selecting the sequence B–NB–NR, which means a blue marble followed by any non-blue marble followed by a non-red marble. The general form for this sequence is `B-(R|G)-(B|G)`, which can be decomposed into the following sequences:

```
B-R-B
B-R-G
B-G-B
B-G-G
```

The preceding sequences have the following probabilities:

```
5/10*3/9*4/8
5/10*3/9*2/8
5/10*2/9*4/8
5/10*2/9*1/8
```

The sum of the preceding four terms can be calculated as follows:

```
5/10*[3/9*4/8+3/9*2/8+2/9*4/8+2/9*1/8]
= 5/10*[3/9*(4/8+2/8)+2/9*(4/8+1/8)]
= 5/10*[3/9*(6/8)+2/9*(5/8)]
= 5/10*[(3*6+2*5)/(9*8)]
= 5/10*[28/72]
= 5/10*[7/18]
= 1/2*[7/18]
= 7/36
```

The Complement of a Probability

Given two independent events A and B, the probability of A and B occurring equals `P(A)*P(B)`. Moreover, the probability that A and B do *not* occur equals `P(not A) + P(not B)`. This simple relationship is useful for solving certain types of tasks.

For example, suppose that container is filled with 40 balls: 10 red, 10 blue, 10 green, and 10 yellow. Furthermore, all balls of each color are numbered 1 through 10.

What is the probability that two balls selected without replacement have the *same* color? Select one ball, after which the container has 39 balls, 9 of which have the same color as the selected ball. Hence, the answer is 9/39.

Similarly, the probability that two cards selected without replacement have the same number is 3/39: there are only four cards with the same number, and after the first card is selected, there are only three remaining cards with the same number.

The final question has multiple parts.

Part 1: What is the probability that two balls that are selected without replacement do *not* have the same color? The answer is 30/39, because 9 remaining balls have the same color.

Part 2: What is the probability that two balls that are selected without replacement do not have the same color *and* they do not have the same number? The first probability is 30/39 and the second probability is 3/39. However, the latter probability is included in the first probability, so we must *subtract* the second probability from the first, so the answer equals `30/39 - 3/39 = 27/39`.

Part 3: What is the probability that two balls that are selected without replacement either have the same color *or* they have the same number? The first probability is 3/39, and the second probability is 9/39. Hence, the answer is `3/39 + 9/39 = 12/39`. Notice that the value `12/39 = 1 - 27/39 =` the complement of the task in Part 2.

Selecting a Matching Pair

Suppose that a container is filled with 10 red marbles and 10 green marbles. How many times must you retrieve a marble (without replacement) to *guarantee* that there are two marbles of the same color?

Despite its appearance, this task is *not* a probability-based question. Let's draw one marble at a time and look at the sequence of colors. If we select the first marble, we can have either of the following:

```
R
G
```

The preceding possibilities obviously do not contain two marbles of the same color. Now select the second marble and we can have the following sequences, where the left-most value is either R or G:

```
RR
RG
GR
GG
```

There are two matching pairs, as well as two pairs that do not match, so once again, selecting two marbles does not guarantee that every sequence contains a pair of matching colors. Hence, we need to select a third marble, and we have the following sequences:

```
RRR
RRG
RGR
RGG
RRR
RRG
RGR
RGG
```

Each of the preceding set of 8 sequences *does* contain a pair of matching colors, so the answer is 3.

CHILDREN-RELATED PROBABILITIES

There are four possible ways in which a couple can have two children (either through birth, adoption, or some combination):

- BB
- BG
- GB
- GG

Note that the preceding combinations treat the order of birth as distinct, and therefore BG and GB are considered distinct. Hence, we have the following probabilities:

- a pair of boys (1/4)
- one boy and one girl (1/2)
- a pair of girls (1/4)

Now suppose that a couple has one boy: then the probability that the next child is a boy equals 1/2 because the second child can be either a boy or a girl, and each outcome has the same probability. Note that this scenario assumes that the gender of a child is independent of the gender of any previous child (which might not always be true in real life).

However, consider the case in which a couple has one boy: what is the probability that the couple will have two boys? This question involves conditional probability, and the possibilities are listed here:

* BB
* BG
* GB

There are three elements in the set of possible outcomes because the outcome GG is excluded from this set (the couple already has a boy). Hence, the answer is 1/3 instead of 1/4.

SUMMARY

This chapter started with a discussion of set theory, open and closed sets, and then an explanation of Brouer's fixed point theorem. Next, you learned about basic concepts in probability, such as how to calculate the expected value of a set of numbers (with associated probabilities), the concept of a random variable (discrete and continuous), and a short list of some well-known probability distributions. In addition, you learned how to solve coin tossing tasks and dice tossing tasks, followed by card drawing tasks and bucket-related tasks that are performed without replacement.

Moreover, you learned about recursion for calculating quantities such as the factorial value of a positive integer and Fibonacci numbers. In addition, you learned how to use recursion to calculate the sum of an arithmetic series of numbers as well as the sum of a geometric series of numbers. Finally, you learned about basic combinatorial concepts, such as calculating combinations and permutations of a set of objects.

CHAPTER 3

INTRODUCTION TO STATISTICS

This chapter introduces concepts in statistics, such as the mean, median, mode, and standard deviation. You will also learn about random variables and two important results in probability, which are the Law of Large Numbers and the Central Limit Theorem.

The first section of this chapter briefly describes the need for statistics, followed by some well-known terminology in statistics. This section also introduces the content of random variable, as well as the difference between discrete and continuous random variables.

The second section introduces basic statistical concepts, such as *mean*, *median*, *mode*, *variance*, and *standard deviation*, along with Python examples that show you how to calculate these quantities. You will also learn about Chebyshev's inequality, dispersion, and sampling techniques.

INTRODUCTION TO STATISTICS

Statistics are useful for describing the characteristics of a dataset, which can be a sample of a population. A *population* is a set or a collection of entities, where the latter can be people, inanimate objects, or abstractions of a physical entity. A *sample* is a subset of the population that is selected for study.

A *paired sample* (also known as a matched or dependent sample) refers to two sets of data where the observations in one set are paired or matched with observations in the other set. This pairing is typically based on some intrinsic connection or relationship between the observations. An *independent sample* involves two samples whose values are from two different populations.

A *variable* is a characteristic that can have different values. A *quantitative variable* has values that are real numbers, whereas a *qualitative variable* consists of categorical data (i.e., string values). The value of a variable is called an *observation* or *measurement*.

A *discrete variable* is a variable that is countable in the mathematical sense. Hence, a discrete variable can consist of a finite set of values as well as a set of values that can be "mapped" to the set of positive integers.

A *continuous variable* consists of an uncountably infinite set of values, such as the real number line. Note that irrational values must be approximated, which effectively means that the values of a continuous variable are treated as rational values.

Inferential Statistics Versus Descriptive Statistics

Descriptive statistics are useful when you need a summary of the nature of the data in a dataset, which includes the mean, median, and standard deviation.

Inferential statistics involves making inferences from a sample of data from the population. This technique is used in various scenarios, such as predicting voter outcomes in elections or checking trucks for food spoilage. This approach is more cost effective when a population is so large that it is impractical to validate every item in the population.

Various types of descriptive statistics that you can obtain using ordinal data are listed here:

- frequency distribution
- measures of central tendency: mode and median
- measures of variability: range

A *frequency distribution* pertains to the distribution of data in a dataset. You can summarize the distribution via a frequency table, a histogram, or a pivot table (among others). If possible, try to display the data in a dataset in a histogram, which might provide some clues regarding the distribution of the data.

BASIC CONCEPTS IN STATISTICS

This section discusses the mean, median, and mode, followed by a section that discusses variance and standard deviation. Feel free to skim (or skip) this section if you are already familiar with these concepts. Here is a summary of different types of measures in statistics:

- measures of central tendency: the mean, median, mode, and midrange
- measures of variation: the range, variance, and standard deviation
- percentiles, deciles, and quartiles: finding the relative location of a data point in a dataset
- For the following subsections, let's suppose that the set X ={x1, ..., xn} is a set of numbers that can be positive, negative, integer-valued, or decimal values.

The Mean

The *mean* of the numbers in the set X is the average of the values. For example, if the set X consists of {-10,35,75,100}, then the mean equals (-10 + 35 + 75 + 100)/4 = 50. If the set X consists of {2,2,2,2}, then the mean equals (2+2+2+2)/4 = 2. As you can see, the mean value is not necessarily one of the values in the set.

Keep in mind that the mean is sensitive to outliers. For example, the mean of the set of numbers {1,2,3,4} is 2.5, whereas the mean of the set of number {1,2,3,4,1000} is 202. Since the formulas for the variance and standard deviation involve the mean of a set of numbers, both of these terms are also more sensitive to outliers. The mean is suitable for symmetric distributions without outliers.

The Weighted Mean

The weighted mean calculates a sum of a set of values where some (possibly all) values in the set are multiplied by different weights. The weighted sum is calculated in the same manner as the expected value. For example, suppose that we have the sets V and W as shown here:

```
V = {1, 2, 3, 4}
W = {0.5, 2, 1.5, -2}
```

Then the weighted average WA is calculated as follows:

```
      4
WA = SUM vi * wi)/4
     i=1
```

Hence, the weighted sum WA equals the following value:

```
WA = [1*0.5 + 2*2 + 3*1.5 +4*(-2)]/4 = (0.5+4+4.5-8)/4
   = 1/4
```

The Median

The *median* of the numbers (sorted in increasing or decreasing order) in the set X is the middle value in the set of values, which means that half the numbers in the set are less than the median and half the numbers in the set are greater than the median. Note that the median is not necessarily one of the values in the given set.

For example, if the set X consists of {35,75,100}, then the *median* equals 75 because this value partitions the set X into two sets having the same cardinality. However, if the set X consists of {-10,35,75,100}, then the *median* equals 55 because 55 is the average of the two numbers 35 and 75. As you can see, half the numbers are less than 55 and half the numbers are greater than 55. If the set X consists of {2,2,2,2}, then the *median* equals 2.

By contrast, the median is much less sensitive to outliers than the mean. For example, the median of the set of numbers {1,2,3,4} is 2.5, and the median of the set of numbers {1,2,3,4,1000} is 3.

As you can see from the preceding simple example, the median works better than the mean for skewed distributions or data with outliers.

The Mean Versus the Median

The preceding section stated that the median works better than the mean for skewed distributions or data with outliers, which is true for many situations. However, there are cases in which the mean is preferred instead of the median, even with a skewed dataset. Specifically, we prefer the median when we want to *avoid* an outlier, whereas we prefer the mean when we *want* the outlier.

For example, suppose that the values in set A consist of {$2,$4,$5,$6,$8}, so the mean and median are both $5. In addition, suppose that set B contains the values {$2,$3,$4,$6,$1000000}, so the mean is $200,003 and median is $4. Hence, it is reasonable to use the mean for set A and the median for set B: this is a logical choice, because 1,000,000 in set B appears to be an outlier with respect to the other values in set B.

Now let's assume that the values in A and B represent monetary rewards, where all outcomes in A and in B have equal probability (i.e., 1/5). Then the potential payoff for set A is much smaller than the potential payoff for set B. In the former case, the maximum you can receive is $8, whereas in the latter case, you might receive $1,000,000.

As you can see, set A is close to symmetric with median equal to 5, whereas set B is highly asymmetric whose median is smaller than the mean.

In essence, we *prefer* set B because it contains an outlier that can lead to a positive outcome.

The Mode

The *mode* of the numbers (sorted in increasing or decreasing order) in the set X is the most frequently occurring value, which means that there can be more than one such value. The median is frequently the preferred statistic for skewed distributions or distributions with outliers.

As a simple example, if the set X consists of {2,2,2,2}, then the *mode* equals 2. However, if X is the set of numbers {2,4,5,5,6,8}, then the number 5 occurs twice and the other numbers occur only once, so the *mode* equals 5.

If X is the set of numbers {2,2,4,5,5,6,8}, then the numbers 2 and 5 occur twice and the other numbers occur only once, so the *mode* equals 2 and 5. A set that has two modes is called *bimodal*, and a set that has more than two modes is called *multimodal*.

One other scenario involves sets that have numbers with the same frequency and they are all different. In this case, the mode does not provide meaningful information, and one alternative is to partition the numbers into subsets and then select the largest subset. For example, if set X has the values {1,2,15,16,17,25,35,50}, we can partition the set into subsets whose elements are in range that are multiples of ten, which results in the subsets {1,2}, {15,16,17}, {25}, {35}, and {50}. The largest subset is {15,16,17}, so we could select the number 16 as the mode.

As another example, if set X has the values {-10,35,75,100}, then partitioning this set does not provide any additional information, so it is probably better to work with either the mean or the median.

Calculating Interquartile Values

This section shows you how to calculate Q2, followed by Q1, and finally Q3 for a discrete set of integer values. Let's suppose that the set X consists of the values {5, 6, 7, 9, 10, 12, 14}.

The second quartile Q2 is the median value of X, which equals 9. The first quartile Q1 is the median of the data values of X that are less than the second quartile Q2 (i.e., 9), which in turn equals 6. The third quartile Q1 is the median of the data values of X that are greater than the second quartile Q2 (i.e., 9), which in turn equals 12. Therefore, the interquartile term IQR equals Q3-Q1, which equals 6. In tabular form, here are the calculated results:

```
Q2 = 9
Q1 = 6
Q3 = 12
IQR = 6
```

A Code Sample with the Mean, Median, and Mode

Listing 3.1 displays the content of the Python file mean_median_mode. py that shows you how to calculate the mean, median, and mode of a set of numbers.

Listing 3.1: mean_median_mode.py

```python
import statistics as stats

data1 = [1,2,3,4,5]
mean   = stats.mean(data1)
median = stats.median(data1)
mode   = stats.mode(data1)

print("data1: ",data1)
print("mean:   ",mean)
print("median:",median)
print("mode:   ",mode)
print()

data2 = [1,1,1,2,2,2,3,4,5,20]
mean   = stats.mean(data2)
median = stats.median(data2)
mode   = stats.mode(data2)

print("data2: ",data2)
print("mean:   ",mean)
print("median:",median)
print("mode:   ",mode)
```

Listing 3.1 starts with an `import` statement, followed by the variable `data1` that is initialized as a list of integers. The next three code snippets calculate the mean, median, and mode of the values in `data1`, and then display the results.

The second block of code is similar to the previous code block, using the variable `data2` instead of `data1`. Launch the code in Listing 3.1, and you will see the following output:

```
data1:  [1, 2, 3, 4, 5]
mean:   3
median: 3
mode:   1

data2:  [1, 1, 1, 2, 2, 2, 3, 4, 5, 20]
mean:   4.1
median: 2.0
mode:   1

data3:  [-2, -1, 1, 2, 30, 40, 50, 2000]
mean:   265
median: 16.0
mode:   -2
```

Here are suggestions for the use of the mean, median, and mode:

- mean: not skewed interval/ratio
- median: ordinal or skewed interval/ratio
- mode: nominal data

The Arithmetic Mean, Harmonic Mean, and Geometric Mean

The *arithmetic mean* is the average value of given data points, which is calculated by dividing the sum of all the observations by the total number of observations. For example, if `a` and `b` are two real numbers, then the arithmetic mean is `(a+b)/2`.

The *harmonic mean* is more complex: it equals the reciprocal of the arithmetic mean of reciprocal values. Given the real numbers `a` and `b`, the harmonic mean equals `2/[1/a + 1/b]`. Incidentally, the term "harmonic mean" is influenced by the *harmonic series*, which is the sum of the reciprocal of all positive integers: `1/1 + 1/2 + 1/3 +...` Although it is not obvious, the sum of this sequence of numbers is infinite (check online for a proof).

The *geometric mean* of two numbers is the square root of the product of those numbers (provided that the product is nonnegative). This can be easily generalized: if the set of `k` numbers `{n1, n2, . . . , nk}` has a positive product, then the geometric mean of those numbers is the `k`th root of `n1*n2*...*nk`.

As a simple example, suppose that `a = 4` and `b = 16` (conveniently chosen for the following calculations), then the average of `a` and `b = (4+16)/2 = 10`; the harmonic mean of `a` and `b = 1/(1/4 + 16) = 1/(5/16) = 16/5`; and the geometric mean of `a` and `b = sqrt(4*16) = sqrt(64) = 8`.

THE VARIANCE AND STANDARD DEVIATION

The *variance* of a distribution is `E((X_bar - X)**2)`, which is the mean of the squared difference from the mean. Hence, the variance measures the variability of the numbers from the average value of that same set of numbers. The *standard deviation* is the square root of the variance.

Another way to describe the *variance* is the sum of the squares of the difference between the numbers in `X` and the mean `mu` of the set `X`, divided by the number of values in `X`, as shown here:

```
           n
variance = [SUM (xi - mu)**2 ] / n
          i=1
```

For example, if the set `X` consists of `{-10,35,75,100}`, then the *mean* equals `(-10 + 35 + 75 + 100)/4 = 50`, and the variance is computed as follows:

```
variance = [(-10-50)**2 + (35-50)**2 + (75-50)**2 +
(100-50)**2]/4
         = [60**2 + 15**2 + 25**2 + 50**2]/4
         = [3600 + 225 + 625 + 2500]/4
         = 6950/4 = 1,737
```

The standard deviation `std` is the square root of the variance, as shown here:

```
std = sqrt(1737) = 41.677
```

If the set X consists of {2,2,2,2}, then the *mean* equals (2+2+2+2)/4 = 2, and the variance is computed as follows:

```
variance = [(2-2)**2 + (2-2)**2 + (2-2)**2 + (2-
2)**2]/4
         = [0**2 + 0**2 + 0**2 + 0**2]/4
         = 0
```

The standard deviation std is the square root of the variance:

```
std = sqrt(0) = 0
```

Standard Deviation for a Sample Versus a Population

The standard deviation for a sample and the standard deviation for a population contain the quantity n-1 and n, respectively, in the denominator. The reason for this slight difference is that the value n underestimates the true value of the variance and (therefore standard deviation) in the population, which results in a *biased* estimate.

The solution involves replacing n by (n-1) in the denominator to achieve an unbiased estimate. By contrast, notice that the mean of a sample and the mean of a population both contain the quantity n in the denominator.

A Simple Code Sample

Listing 3.2 displays the content of the Python file var_pvar_stdev.py that shows you how to calculate the variance, pvariance, and standard deviation stdev of a set of numbers.

Listing 3.2: var_pvar_stdev.py

```
import numpy as np
import statistics as stats

values = np.array([1,2,3,4,5,20])
variance  = stats.variance(values)
pvariance = stats.pvariance(values)
stdev     = stats.stdev(values)

print("values:    ",values)
print("pvariance:",pvariance)
```

```
print("variance: ",variance)
print("stdev:    ",stdev)
```

Listing 3.2 starts with two `import` statements, followed by a block of code that computes the `variance`, `pvariance`, and `stdev` of the numbers in the NumPy array `values`. The next block of code displays the values that were calculated in the first block of code. Launch the code in Listing 3.2, and you will see the following output:

```
values:    [ 1  2  3  4  5 20]
pvariance: 41
variance:  50
stdev:     7.0710678118654755
```

Chebyshev's Inequality

Chebyshev's inequality provides a simple way to determine the minimum percentage of data that lies within `k` standard deviations. Specifically, this inequality states that for any positive integer `k` greater than 1, the amount of data in a sample that lies within `k` standard deviations is at least `1 - 1/k**2`. This formula is plausible, because larger values of `k` encompass a larger set of data values, and the quantity `1 - 1/k**2` increases quadratically as the value of `k` increases linearly. For example, if `k = 2`, then at least `1 - 1/2**2 = 3/4` of the data must lie within 2 standard deviations; if `k = 3`, then at least `1 - 1/3**2 = 8/9` of the data must lie within 3 standard deviations.

The interesting part of the term `1 - 1/k**2` is that it has been *mathematically proven* to be true; i.e., it is not an empirical or heuristic-based result. An extensive description regarding Chebyshev's inequality (including some advanced mathematical explanations) is online:

https://en.wikipedia.org/wiki/Chebyshev%27s_inequality

THE MOMENTS OF A FUNCTION (OPTIONAL)

The previous sections describe several statistical terms that are sufficient for understanding the material in this book. However, several of those terms can be viewed from the perspective of different moments of a function.

In brief, the moments of a function are measures that provide information regarding the shape of the graph of a function. In the case of a probability distribution, the first four moments are defined as follows:

- The mean is the first central moment.
- The variance is the second central moment.
- The skewness is the third central moment.
- The kurtosis is the fourth central moment.

More detailed information (including the relevant integrals) regarding moments of a function is available online:

https://en.wikipedia.org/wiki/Moment_(mathematics)#Variance

What is Skewness?

Skewness is a measure of the asymmetry of a probability distribution. A Gaussian distribution is symmetric, which means that its skew value is zero. In addition, the skewness of a distribution is the *third* moment of the distribution.

A distribution can be skewed on the left side or on the right side. A *left-sided* skew means that the long tail is on the left side of the curve, with the following relationships:

```
mean < median < mode
```

A *right-sided* skew means that the long tail is on the right side of the curve, with the following relationships (compare with the left-sided skew):

```
mode < median < mean
```

If need be, you can transform skewed data to a normally distributed dataset using one of the following techniques (which depends on the specific use-case):

- exponential transform
- log transform
- power transform

Perform an online search for more information regarding the preceding transforms and when to use each of these transforms.

What is Kurtosis?

Kurtosis is related to the skewness of a probability distribution, in the sense that both of them assess the asymmetry of a probability distribution. The kurtosis of a distribution is a scaled version of the *fourth* moment of the distribution, whereas its skewness is the *third* moment of the distribution. Note that the kurtosis of a univariate distribution equals 3.

If you are interested in learning about additional kurtosis-related concepts, you can perform an online search for information regarding mesokurtic, leptokurtic, and platykurtic types of "excess kurtosis."

A Code Sample for Skewness and Kurtosis

Listing 3.3 displays the content of the Python file `skew_kurtosis.py` that shows you how to calculate the skew and kurtosis of a set of numbers.

Listing 3.3: skew_kurtosis.py

```
import numpy as np
from scipy.stats import kurtosis, skew
import matplotlib.pyplot as plt

values = np.random.normal(-4,4,1000)
plt.hist(values,bins=40)
print("skew:    ",skew(values))
print("kurtosis:",kurtosis(values))
plt.show()
```

Listing 3.3 starts with `import` statements and then initializes the variable `values` as a NumPy array of 1,000 random values in the interval `[-4,4]`. The next portion of Listing 3.3 plots the random values in a histogram, and also displays the skew and kurtosis of the variable values. Launch the code in Listing 3.3, and you will see the following output:

```
skew:     -0.05308152055274579
kurtosis: -0.21205538639216437
```

Figure 3.1 displays a histogram with 40 bins of the numbers in the `values` array in Listing 3.3.

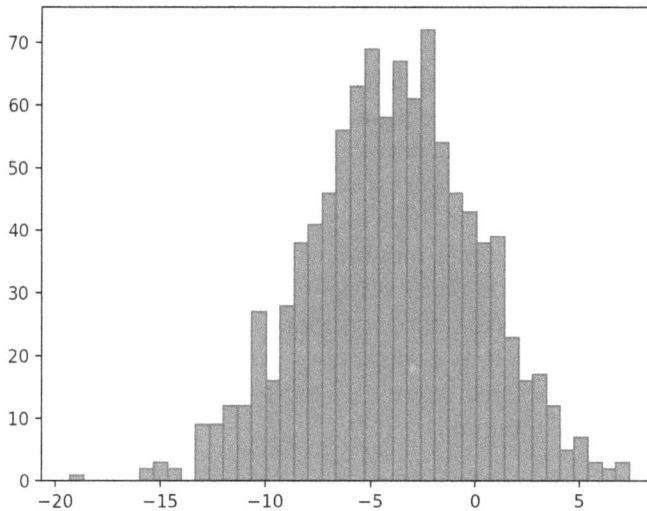

FIGURE 3.1: A histogram of data points

RANDOM VARIABLES

A *random variable* is a variable that can have multiple values, and where each value has an associated probability of occurrence. For example, if X is a random variable whose values are the outcomes of tossing a well-balanced die, then the values of X are the numbers in the set {1,2,3,4,5,6}. Moreover, each of those values can occur with equal probability (which is 1/6).

In the case of two well-balanced dice, let X be a random variable whose values can be any of the numbers in the set {2,3,4, . . . , 12}. The set of associated probabilities for the different values for X are listed here:

{1/36,2/36,3/36,4/36,5/36,6/36,5/36,4/36,3/26,2/36,1/3 6}

Discrete Versus Continuous Random Variables

The preceding section contains examples of *discrete* random variables because the list of possible values is either finite or countably infinite (such as the set of integers). As an aside, the set of rational numbers is also countably infinite, but the set of irrational numbers and also the set of real numbers are both uncountably infinite (proofs are available online).

The set of probabilities of a random variable must form a probability distribution, which means that the probability values are non-negative. In the case of a discrete probability distribution, the sum of the probabilities must equal 1, whereas for a continuous probability distribution, the area under the curve (i.e., between the curve and the horizontal axis) must equal 1.

A *continuous* random variable is a random variable whose values can be *any* number in a given interval such as [a,b] (where a<b), which contains an uncountably infinite number of values. For example, the amount of time required to perform a task is represented by a continuous random variable.

Moreover, a continuous random variable has a probability distribution that is represented as a continuous function. The constraint for such a variable is that the area under the curve equals 1 (sometimes calculated via a mathematical integral).

Confounding Variables

A *confounding variable* (also called a *confounding factor*) is a variable that influences the dependent variable and the independent variable, which creates a spurious association that can lead to erroneous conclusions.

A famous example is the observation that an increase in ice cream consumption during the summer in a city coincided with an increase in crime. We would not assert that ice cream consumption leads to greater crime: temperature is an underlying confounding variable.

Interestingly, one recent study suggests that "the seemingly protective effect of vegetable intake against cardiovascular disease risk is very likely to be accounted for by bias from residual confounding factors, related to differences in socioeconomic situation and lifestyle." More information can be found online:

https://www.yahoo.com/news/why-packing-diet-vegetables-may-205810092.html

An explanation of a counterfactual is available online:

https://ml-retrospectives.github.io/neurips2020/camera_ready/5.pdf

Counterfactual Analysis

Randomized controlled trials (RCTs) establish causality between variables via counterfactual analysis. Examples of RCTs include drug and vaccine testing. RCTs were instrumental in analyzing the impact of smoking as a cause of cancer.

However, the ability of RCTs to determine causality is only possible for a small number of variables, and the experiments themselves are also challenging to set up.

Interestingly, AirBnB developed a new technique called Artificial Counterfactual Estimation (ACE) that reproduces counterfactual outcomes by a combination of machine learning techniques and causal inference. More information and a description of ACE are available online:

https://medium.com/airbnb-engineering/artificial-counterfactual-estimation-ace-machine-learning-based-causal-inference-at-airbnb-ee32ee4d0512

MULTIPLE RANDOM VARIABLES

This section briefly discusses some basic properties (without derivations) of multiple random variables. If $X1$ and $X2$ are two random variables and $Y = X1+X2$, then the following formulas are valid:

```
mean(Y) = mean(X1) + mean(X2)

var(Y)  = var(X1)  + var(X2) + 2*Covariance(X1,X2)
```

For the general case involving a sequence of random variables $X1$, $X2$, $X3$, $\ldots$, Xn, we have the following formulas:

```
Y = X1 + X2 + ... + Xn
             n
mean(Y) = SUM mean(Xi)
          i=0


             n
var(Y) = SUM var (Xi) + 2*SUM Covariance(Xi,Xj)
         i=0                 i<j
```

If Xi and Xj are pairwise independent, then `covariance(Xi,Xj)` = 0, and therefore we have the following:

```
            n
var(Y)  = SUM var (Xi)
          i=0
```

Working with Two or More Variables

There are several constructs that you can use when you need to work with two or more variables, some of which are listed here:

- contingency table (useful for two categorical variables)
- contour plot
- hexagonal binning
- violin plot

Types of Convergence for Random Variables

Consider a sequence of random variables X1, X2, X3, ... that can exhibit the following types of convergence, shown in increasing order:

- convergence in mean
- convergence in probability
- convergence in distribution

"Increasing order" indicates the following: convergence in mean (the strongest) implies convergence in probability, which in turn implies convergence in distribution (the weakest).

Using informal terminology, a sequence {X1, X2, ..., Xn} "convergences in mean" to X indicates that the difference |X - Xn| becomes arbitrarily small as n becomes arbitrarily large. Similar comments apply to convergence in probability and distribution.

SAMPLING TECHNIQUES FOR A POPULATION

This section contains sampling techniques that pertain to creating a sample from a given population. Note that the word *population* specifically refers to the entire set of entities in a given group, such as the population of a country, the people over 65 in the USA, or the number of first year students in a university.

However, in many cases, statistical quantities are calculated using samples instead of an entire population. Thus, a sample is (a much smaller) subset of the given population. (See the section on the Central Limit Theorem regarding the distribution of the mean of a set of random samples of a population, which need not be a population with a Gaussian distribution.)

The following list of items contains common techniques for sampling data from a population:

- random sampling
- cluster sampling
- convenience sampling
- quota sampling
- simple random sampling
- stratified sampling
- systematic sampling

One other important point to consider is that the population variance is calculated by multiplying the sample variance by $n/(n-1)$, as shown here:

```
population variance = [n/(n-1)]*variance
```

Random sampling (also referred to as the *probability sampling*) involves random samples of observations from a population, where samples have an equal probability of being selected. This method reduces the bias in the sample.

Systematic sampling involves selecting samples in a "periodic" fashion. For example, suppose that you want to form three teams from a roomful of people. Next, starting from a convenient location in the room (e.g., front left corner or rear right corner), the first three adjacent people will say "one," "two," and "three," respectively, and the process will continue in this manner until everyone has designated one of those three numbers.

Stratified sampling involves subdividing a population into subgroups (called *strata*), after which a random sample is selected from each subgroup. The selection process continues until the sample contains the specified number of samples.

Cluster sampling involves the following steps:

1. Subdivide a population into subgroups ("clusters").

2. Randomly select some of the subgroups.

3. Randomly select samples from the chosen subgroups.

Hence, cluster sampling involves randomly selecting people from each cluster instead of selecting people from the entire group. Note that if you specify only one cluster, then cluster sampling is the same as random sampling.

WHAT IS BIAS?

In the context of machine learning and data science, *bias* in a dataset refers to certain systematic errors or prejudices that affect the representativeness and fairness of the data. This can occur at various stages of data collection, processing, and analysis. Here are some common types of bias in datasets, followed by brief explanations:

- sampling bias
- measurement bias
- selection bias
- confirmation bias
- label bias
- class imbalance
- time bias
- geographical bias
- cultural bias

Sampling bias occurs when the data collected is not representative of the population it is meant to represent. For example, an online survey might only represent the views of those with Internet access.

Measurement bias happens when data is consistently and systematically collected inaccurately, often due to faulty measurement instruments or human error.

Selection bias occurs when the data used for analysis is not randomly selected from the population, leading to results that cannot be generalized. For example, studying the effects of a drug only on volunteers may not be representative of its effects on the broader population.

Confirmation bias occurs when data is selectively collected or chosen to confirm a pre-existing belief or hypothesis, thereby skewing the results.

Label bias in supervised learning occurs when the labels used for training the model are systematically incorrect or biased, ensuring that the model's predictions will also be biased.

Class imbalance in classification problems occurs if one class is significantly underrepresented in the training data; the model may be biased toward the majority class.

Time bias occurs when data collected at different times is influenced by external factors like seasonality or economic conditions, leading to biased results.

Geographical bias occurs when the data is not representative of all geographical locations it is supposed to cover, leading to biased conclusions.

Cultural bias occurs when the data reflects prejudices or cultural beliefs. It can lead to models that reinforce these biases.

Impact of Bias

Bias can have an impact in several ways, some of which are listed here:

- model performance
- fairness
- decision-making

Model performance bias in the training data can lead to poor model generalization and misleading insights. A biased model can perpetuate or even exacerbate existing social biases, leading to unfair or unethical outcomes. Biased data can lead to flawed decision-making in various applications like healthcare, finance, and criminal justice.

TWO IMPORTANT RESULTS IN PROBABILITY

In addition to the concepts that you have learned so far in this chapter, there are several other important results that you will encounter if you decide to pursue statistics in greater depth.

The Law of Large Numbers

The *Law of Large Numbers* (LLN) states that as the size of a sample set from a population increases, the accuracy of an estimate the sample mean will be of the population mean will also increase.

The LLN is particularly important for long-term results. As you can probably surmise, the LLN is important for casinos because they are more interested in the long-term outcome of bets by customers than short-term outcomes.

Other examples of the LLN pertain to flipping a balanced coin: approximately 50% of the outcomes will be heads and 50% of the outcomes will be tails. Similar results are true for tossing a six-sided die: each of the six outcomes has an equal probability, which means that each outcome will occur approximately $1/6^{th}$ of the number of outcomes for a large number of outcomes.

More detailed information regarding the LLN is accessible online:

https://en.wikipedia.org/wiki/Law_of_large_numbers

Another interesting result is the *Law of Truly Large Numbers*, and you can read about some interesting results here:

https://en.wikipedia.org/wiki/Law_of_truly_large_numbers

The Central Limit Theorem

The *Central Limit Theorem* (CLT) states that the distribution of mean values of a set of independent and identically distributed random values tend toward the standard normal distribution, even if the original variables are not normally distributed.

The CLT is a remarkable result because the statistics that are used for normal distributions can be used for many other types of distributions. Moreover, the *De Moivre-Laplace Theorem*, which is a special case of the CLT, states that a normal distribution can be used as an approximation for a binary distribution under certain conditions. Details regarding the De Moivre-Laplace Theorem are accessible online:

https://en.wikipedia.org/wiki/De_Moivre%E2%80%93Laplace_theorem

SUMMARY

This chapter started with a brief description regarding the need for statistics, followed by some well-known concepts, such as the mean, median, mode, variance, and standard deviation. In addition, you learned about Chebyshev's inequality, dispersion, and sampling techniques.

Moreover, you learned about the concept of a random variable, which can be either continuous or discrete. Finally, you learned about the Law of Large Numbers and the Central Limit Theorem, both of which are extremely important in probability as well as statistics.

## CHAPTER 4

METRICS IN STATISTICS

This chapter introduces the confusion matrix that you can generate via classification algorithms for machine learning. You will also learn about various metrics for categorical data, such as accuracy, recall, and precision. In addition, you will learn about metrics for continuous data, such as R^2 (R^2) and adjusted R^2 (adjusted R^2).

The first section introduces the confusion matrix and concepts such as true positive, true negative, false positive, and false negative. In addition, you will learn about Type I and Type II errors, as well as accuracy and balanced accuracy, along with a caveat regarding the accuracy metric.

The second section discusses recall, precision, specificity, and prevalence. You will learn about the difference between precision and recall, and how to decide which one is more relevant. Furthermore, you will learn about the terms TPR, FPR, PV, FDR, and FOR.

The third section introduces statistics for continuous data, such as RSS, TSS, and R^2. You will learn about adjusted R^2 and the limitations of R^2. The fourth section discusses MAE, MSE, RMSE, and MRR, along with a code sample that contains these terms.

The fifth section introduces statistics for categorical data, such as the F1 score and related scores, such as F2, and F3 that are useful for classification algorithms. Finally, you will learn about skewness and kurtosis.

THE CONFUSION MATRIX

A *confusion matrix* provides information that is well-suited for classification tasks: it shows you how many observations were classified by a model. Specifically, a confusion matrix contains the number of observations that are correctly classified as well as the number of observations that are incorrectly satisfied.

Classification algorithms in machine learning that involve code from the scikit-learn library enable you to generate a confusion matrix. Fortunately, you can learn about the contents of a confusion matrix without any knowledge of machine learning and classification models. Thus, you can perform an analysis of the efficacy of a given algorithm with a given dataset. In the case of two classes (e.g., "healthy" is considered "negative" whereas "sick" is considered "positive"), there are four outcomes:

- true positive (TP)
- false positive (FP)
- true negative (TN)
- false negative (FN)

A *true positive* equals the number of values that are *correctly* identified as positive, whereas a *false positive* equals the number of values that are *incorrectly* identified as positive. In an analogous manner, a *true negative* equals the number of values that are *correctly* identified as negative, whereas a *false negative* equals the number of values that are *incorrectly* identified as negative. Moreover, the preceding four quantities occupy the four cells of the following 2x2 confusion matrix:

```
TP | FP
-------
FN | TN
```

An example of a confusion matrix with numeric values is shown here, followed by the interpretation of the values in the confusion matrix:

```
[[64  4]
 [ 3 29]]
```

The four values in the preceding 2x2 matrix represent the following quantities:

```
TP = True positive: 64
FP = False positive: 4
TN = True negative: 29
FN = False negative: 3
```

Another example of a confusion matrix involves three outcomes, which means that the confusion matrix is 3x3 instead of 2x2:

```
[[12  0  2]
 [ 0 15  1]
 [ 2  0  4]]
```

In addition to 2x2 and 3x3 confusion matrices, a `nxn` confusion matrix is generated when a feature consists of `n` labels for a given class.

As a practical example, suppose that a dataset that contains clinical trial data for cancer, which involves two classes (healthy and sick). Once again, there are four possible outcomes: true positive, false positive, true negative, and false negative (discussed later). A confusion matrix contains numeric (integer) values for these four quantities.

Normalized Confusion Matrix

If `cm` is a confusion matrix, such as the confusion matrix in the previous section, the following code snippet normalizes the values in that matrix so that the sum of the values in each row equals 1:

```
cm = np.array([[64, 4],[3, 29]])
normalized_cm = cm.astype('float') / cm.sum(axis=1)[:,
np.newaxis]
```

An even simpler way to normalize the values in a confusion matrix `cm` is shown here, where `y_true` are the actual labels in a dataset and `y_pred` are the predicted values that are compared with the actual labels to generate a confusion matrix:

```
cm(y_true, y_pred, normalize='all')
```

A third way to normalize a confusion matrix involves `scikitplot`, as shown here:

```
import scikitplot as skplt
skplt.metrics.plot_confusion_matrix(Y_TRUE,Y_
PRED,normalize=True)
```

The value for normalization in the preceding code snippet might also depend on the version of Python that you have installed on your machine.

We now use the confusion matrix from the previous code snippet that initializes the variables `cm` and `normalized_cm` to create the corresponding normalized confusion matrix:

```
|0.94117647 0.05882353|
|0.09375    0.90625   |
```

A Code Sample of a Confusion Matrix

Listing 4.1 displays the content of the Python file confusion_matrix.py that shows you how to generate a confusion matrix from a set of numeric data values.

Listing 4.1: confusion_matrix.py

```
import matplotlib.pyplot as plt
import pandas as pd
import seaborn as sns
from sklearn.metrics import confusion_matrix

data = {'y_true': [1, 0, 0, 1, 0, 1, 0, 0, 1, 0, 1, 0],
        'y_pred': [1, 1, 0, 1, 0, 1, 1, 0, 1, 0, 0, 0]}

print("=> Data Values:")
print(data)
print()

df = pd.DataFrame(data, columns=['y_true','y_pred'])
print("=> DataFrame df:")
print(df)
print()

cm = pd.crosstab(df['y_true'], df['y_pred'],
rownames=['Actual'], colnames=['Predicted'])
print ("=> Confusion matrix:")
print (cm)
print()

cm2 = confusion_matrix(data['y_true'], data['y_pred'],
normalize='all')
print ("=> Normalized confusion matrix, where the sum
of the values in each row equals 1:")
print (cm2)
```

```
sns.heatmap(cm2, annot=True)
plt.show()
```

Listing 4.1 starts with `import` statements and then initializes the variable `data` with a set of 0s and 1s for the `y_true` and the `p_pred` elements. These values were arbitrarily selected, so there is no significance to the chosen values (feel free to specify different values).

The next code block initializes the data frame `df` with the values in the variable data, after which the confusion matrix `cm` is generated based on the data values in `df`. The confusion matrix is printed, and then a second normalized confusion matrix `cm2` is created and also printed. The last code snippet generates and then displays a Seaborn heat map based on the contents of the confusion matrix `cm2`. Launch the code in Listing 4.1, and you will see the following output:

```
=> Data Values:
{'y_true': [1, 0, 0, 1, 0, 1, 0, 0, 1, 0, 1, 0],
 'y_pred': [1, 1, 0, 1, 0, 1, 1, 0, 1, 0, 0, 0]}

=> DataFrame df:
     y_true  y_pred
0        1       1
1        0       1
2        0       0
3        1       1
4        0       0
5        1       1
6        0       1
7        0       0
8        1       1
9        0       0
10       1       0
11       0       0

=> Confusion matrix:
Predicted  0  1
Actual
```

```
0              5   2
1              1   4


=> Normalized Confusion matrix:
[[0.41666667 0.16666667]
 [0.08333333 0.33333333]]
```

Figure 4.1 displays the heat map generated via the Seaborn package, using the data from the confusion matrix `cm2`.

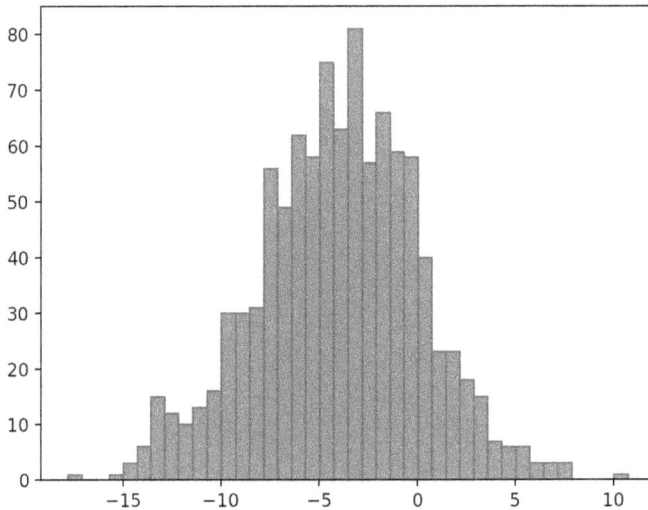

FIGURE 4.1: A best-fitting distribution for a set of random values

What are TP, FP, FN, and TN?

A *binary confusion matrix* (also called an *error matrix*) is a type of contingency table with two rows and two columns that contains the number of false positives, false negatives, true positives, and true negatives. Here is a 2x2 confusion matrix shown again for your convenience:

```
TP | FP
-------
FN | TN
```

The four entries in a 2x2 confusion matrix have labels with the following interpretation:

- TP: true positive
- FP: false positive
- TN: true negative
- FN: false negative

The four entries in the confusion matrix can be described as follows:

- true positive (TP): predicted true and actually true
- true negative (TN): predicted false and actually false
- false positive (FP): predicted true and actually false
- false negative (FN): predicted false and actually true

Hence, the values on the main diagonal of the confusion matrix are *correct* predictions, whereas the off-diagonal values are *incorrect* predictions. In general, a lower FP value is better than a lower FN value. For example, an FP indicates that a healthy person was incorrectly diagnosed with a disease, whereas an FN indicates that an unhealthy person was incorrectly diagnosed as healthy.

Keep in mind that the confusion matrix can be a `nxn` matrix and not just a 2x2 matrix. For example, if a class has 5 possible values, then the confusion matrix is a 5x5 matrix, and the numbers on the main diagonal are the "true positive" results.

Type I and Type II Errors

A *type I error* is a false positive, which means that something is erroneously classified as positive when it is negative. On the other hand, a *type II error* is a false negative, which means that something is erroneously classified as negative when it is positive.

For example, a woman who is classified as pregnant even though she is not pregnant is a type I error. By contrast, a woman who is classified as *not* pregnant even though she *is* pregnant is a type II error.

As another example, a person who is classified as having cancer even though that person is healthy is a type I error. By contrast, a person who is classified as healthy even though that person has cancer is a type II error.

Based on the preceding examples, it is clear that type I and type II errors are not symmetric in terms of the consequences of their misclassification. Sometimes, it is a case of life-and-death classification, in which case a true negative ("I'm not sick!") would be the most desirable. Among the four possible outcomes, the sequence of outcomes, from most desirable to least desirable, would arguably be the following:

1. true negative

2. false positive

3. true positive

4. false negative

Although #3 and #4 are both highly undesirable, the third option provides accurate information that enables people to take appropriate action (i.e., obtain medical assistance), whereas the fourth option delays the time at which people can take the necessary precautions, which could result in a more serious state of illness. In addition, people who are erroneously diagnosed with leukemia or cancer (or some other life-threatening disease) might be needlessly subjected to chemotherapy, which has an unpleasant set of consequences.

Accuracy and Balanced Accuracy

You will often see models evaluated via their accuracy, which is defined by the following formula:

```
accuracy  = % of correct predictions
          = (TP + TN) / total cases


balanced accuracy = (recall+specificity)/2
```

The formula for balanced accuracy involves `recall` and `specificity`, both of which are discussed later. Although accuracy can be a useful indicator, accuracy has limited (and perhaps misleading) value for imbalanced datasets. Accuracy can be an unreliable metric because it yields misleading results in unbalanced data sets. Classes with substantially different sizes are assigned equal importance to both false positive and false negative classifications. For example, declaring cancer as benign is worse than incorrectly informing patients that they are suffering from cancer. Unfortunately, accuracy will not differentiate between these two cases.

A Caveat Regarding Accuracy

Accuracy involves the sum of the true positive and true negative values that are on the main diagonal of the confusion matrix, and disregards type I errors and type II errors (on the off diagonal of the confusion matrix). Moreover, data belonging to the majority class tend to be given a true classification, and significantly imbalanced datasets tend to skew results toward the majority class.

As an example, consider a dataset with 1,000 rows in which 1% of the people are sick: hence, 990 people are healthy and 10 people are sick. Now, we train a model to make predictions on this dataset. The no-code solution is to predict that everyone is healthy, which achieves an accuracy of 99%.

The preceding "no-code solution" is obviously unacceptable because it cannot predict *which* people are sick. Instead of accuracy, consider using one or more of the following:

· Matthew's Correlation Coefficient (CCM)
· Cohen's kappa coefficient
· Student's t-test (for normally distributed data)
· Mann-Whitney U test (for non-normally distributed data)

We can also calculate the values for precision, recall, and F1 scores and compare them with the value of accuracy and see how the models react to imbalanced data.

In general, use the accuracy metric when both classes are equally important and 80% are in the majority class.

Recall, Precision, Specificity, and Prevalence

The definitions of recall, precision, and specificity in the case of a 2x2 confusion matrix are given by the following formulas:

```
recall    = % of correctly identified positive cases
          = TP / (TP + FN)


precision = % of correct positive predictions
          = TP / (TP + FP).
specificity = TN/[TN+FP]


prevalence = (TP+FN)/[TP+TN+FP+FN]
```

One way that might help you remember these formulas is to think of their denominators as the sum of the values in columns or rows, as shown here:

```
·   Accuracy  = sum-of-main-diagonal/(sum-of-all-terms)
·   Precision = TP/(sum-of-row-one)
·   Recall    = TP/(sum-of-column-one)
```

- Specificity = TN/(sum-of-column-two)
- False-positive-rate = TN/(sum-of-column-two)
- F1 score = 2/[1/recall + 1/precision]

Recall (also called *sensitivity*) is the proportion of the correctly predicted positive values in the set of actually positively labeled samples: this equals the fraction of the positively labeled samples that were correctly predicted by the model.

The following code snippet shows you how to invoke the `recall_score()` method, which provides a labels parameter for multi-class classification (note that `y_true` and `y_pred` must be initialized for the code to work):

```
from sklearn.metrics import recall_score
recall_score(y_true, y_pred,
labels=[1,2],average='micro')
```

The following code snippet shows you how to invoke the `precision_score()` method, which provides a labels parameter for multi-class classification (note that `y_true` and `y_pred` must be initialized for the code to work):

```
from sklearn.metrics import precision_score
precision_score(y_true, y_pred,
labels=[1,2],average='micro')
```

Another technique that might help you remember how to calculate precision and recall is to notice that

1. both have the same numerator (=TP)

2. the precision denominator is the sum of the first row

3. the recall denominator is the sum of the first column

Thus, we can describe accuracy, recall, precision, and specificity as follows:

- *Accuracy* is the percentage of correctly classified samples of all the samples.
- *Recall* is the percentages of correctly classified positives of all actual positives.
- *Precision* is the percentage of correctly classified positives from all predicted positives.
- *Specificity* is the proportion of negatively labeled samples that were predicted as negative.
- *Prevalence* is a fraction of total population that is labeled positive.

Precision Versus Recall: How to Decide?

Sometimes precision is more important than recall: of the set of cases that were predicted as valid, how many times were they true? If you are predicting books that are suitable for people under 18, you can afford to reject a few books, but cannot afford to accept bad books. If you are predicting thieves in a supermarket, we need a higher value for precision. As you can probably surmise, customer trust will decrease due to false positives.

Just to emphasize what has already been written, *precision* is the proportion of the samples that are actually positive in the set of positively predicted samples, which is expressed informally as

```
precision = (# of correct positive) / (# of predicted
positive)
```

Note: Precision is important when false positives are more important than false negatives, such as spam detection, and you want to minimize FP.

Recall is the proportion of the samples that are actually positive in the set of actual positive samples, which is expressed informally as

recall = (# of *predicted* positive) / (# of *actual* positive)

Note: *Recall* (sensitivity) is important when false negatives are more important than false positives, such as cancer detection, in which case you want to minimize FN.

TPR, FPR, PV, FDR, and FOR

The quantities TPR, FPR, NPV, FDR, and FOR are additional terms that you might encounter, and they are defined in this section.

TPR = true positive rate
TPR = proportion of positively labeled samples that are *correctly* predicted positive
TPR = TP/[TP+FN] = TP/(sum-of-column-one)

FPR = false positive rate
FPR = proportion of negatively labeled samples that are *incorrectly* predicted positive
FPR = FP/[TN+FP] = FP/(sum-of-column-two)

NPV = negative predictive value or NPV

NPV = proportion of negatively labeled samples that are *correctly* predicted negative
```
NPV = TN/[TN+FN] = TN/(sum-of-row-two)
```

FDR = false discovery rate = 1 - PPV = FP/[TP+FP] = FP/
```
(sum-of-row-one)
```

FOR = false omission rate
```
FOR = 1 - NPV = FN/[TN+FN] = FN/(sum-of-row-two)
```

The following list contains the values of the quantities TPR, FPR, NPV, FDR, and FOR in a single list:

- `TPR = TP/(sum-of-column-one)`
- `FPR = FP/(sum-of-column-two)`
- `NPV = TN/(sum-of-row-two)`
- `FDR = FP/(sum-of-row-one)`
- `FOR = FN/(sum-of-row-two)`

In a previous section, you learned about a confusion matrix, and the following output shows you the calculated values for precision, recall, F1-score, and accuracy that can be generated via scikit-learn:

```
from sklearn.metrics import confusion_matrix,
classification_report
# displays values for precision/recall/f1-score/
support:
data = {'y_true': [1, 0, 0, 1, 0, 1, 0, 0, 1, 0, 1, 0],
        'y_pred': [1, 1, 0, 1, 0, 1, 1, 0, 1, 0, 0, 0]}

print(classification_report(data['y_true'], data['y_pred']))
```

	precision	recall	f1-score	support
0	0.83	0.71	0.77	7
1	0.67	0.80	0.73	5
accuracy			0.75	12

macro avg	0.75	0.76	0.75	12
weighted avg	0.76	0.75	0.75	12

THE ROC CURVE AND AUC CURVE

Receiving Operator Characteristic (ROC) curves plot the performance of a model by displaying the FP (false positive) rate on the horizontal axis and the TP (true positive) rate on the vertical axis. Note that the TN (the true negative rate) is also called the *specificity*. The following URL shows an ROC curve and a diagonal line with a slope of 1, both of which are in the first quadrant of the 2D Euclidean plane:

https://en.wikipedia.org/wiki/Receiver_operating_characteristic

The ROC curve provides a visual comparison of classification models that shows the trade-off between the true positive rate and the false positive rate. Both axes have values between 0 and 1: the vertical axis is the true positive rate (TPR) whereas the horizontal axis is the false positive rate (FPR). The ROC curve provides a view of model performance at different threshold values.

The point to remember is that a ROC curve that "hugs" the vertical axis and the upper horizontal line (corresponding to y=1) reflects a better model. Therefore, the goal is to increase TPR while simultaneously maintaining a low FPR; however, both values increase together, so it is a question of the tolerance level for false positives.

Python Code for Generating Metrics

After selecting one class as positive and another class as negative in a binary classification task, the following (partial) code block shows you how to generate a confusion matrix as well as a report that displays values for TP, FP, FN, and TN:

```
# details omitted for the source of the data
# as well as a classification algorithm
# generate the confusion matrix
from sklearn.metrics import confusion_matrix
cm = confusion_matrix(y_test, y_pred)
print("confusion matrix:")
print(cm)
```

```
from sklearn.metrics import confusion_matrix,
classification_report
print(classification_report(y_test, y_pred))
```

As an example, the preceding code block can generate the following type of output (the numbers depend on the dataset):

```
[[64   4]
 [ 3  29]]
```

	precision	recall	f1-score	support
0	0.96	0.94	0.95	68
1	0.88	0.91	0.89	32
accuracy			0.93	100
macro avg	0.92	0.92	0.92	100
weighted avg	0.93	0.93	0.93	100

What is the AUC curve?

The AUC (Area Under a Curve) is the area under an ROC curve between (0,0) and (1,1), which aggregates the performance of the model at all threshold values. In essence, a threshold value is a cutoff point that separates the classes. When a model predicts the probability of an instance belonging to the positive class, this probability is compared against the threshold value to determine the instance's final predicted class.

The choice of a threshold value in a machine learning classification problem can depend on various factors, and the responsibility for setting it can lie with different parties depending on the context. For instance, data scientists or machine learning engineers can specify one or more threshold values.

The area under the ROC curve (ROC-AUC) is a measure of the accuracy of the model. Models closer to the diagonal are less accurate, and the models with perfect accuracy will have an area of 1.0.

As you saw in a previous section, the ROC curve plots points of the form (TPR, FPR). The AUC is the area bounded by the positive vertical axis, ROC, and positive horizontal axis. If the ROC is identical to the dashed diagonal line, the AUC (area under the curve) is 0.5, and therefore the model result is no different from random guessing. However, if the ROC curve "hugs" the axes mentioned in a previous section, then the AUC is much closer to 1.0.

Thus, the AUC is a positive decimal between 1.0 (excellent fit) to 0.5 (random draw). The predictability of a model can be considered "excellent" if the AUC is more than 0.9, and "good" if the AUC is above 0.8.

The best possible value of the AUC is 1 (a perfect classifier), and the worst value is 0 (if all the predictions are wrong). One other point to keep in mind is that the AUC is independent of the classification threshold value.

Calculating ROC AUC Values

The `Python`-based `lazypredict` library is an open-source module that calculates statistics quantities after training a dataset on multiple algorithms, and it is available online:

https://pypi.org/project/lazypredict/

Listing 4.2 displays the content of the Python file `ROC_AUC.py` that generates ROC and AUC scores.

Listing 4.2: ROC_AUC.py

```
# pip3 install lazypredict
#https://www.kdnuggets.com/2019/10/5-classification-
evaluation-metrics-every-data-scientist-must-know.html
import numpy as np
from sklearn.metrics import roc_auc_score

y_true = np.array([0, 0, 0, 1, 1, 1])
y_scores = np.array([0.3, 0.1, 0.4, 0.35, 0.8, 0.5])
scores = roc_auc_score(y_true, y_scores)
print("ROC AUC scores:",scores)
```

Listing 4.2 starts with two `import` statements and then initializes the variables `y_true` and `y_scores` as two NumPy arrays of equal length. Next, the `scores` variable is initialized with the result of invoking the method `roc_auc_score()` with `y_true` and `y_scores`. Launch the code in Listing 4.2, and you will see the following output:

```
ROC AUC scores:  0.888888888888889
```

Listing 4.3 displays the content of the Python file `lazypredict1.py` that shows you how to use the `lazypredict Python` module for calculating values such as ROC AUC and F1 scores for a given dataset.

Listing 4.3: lazypredict1.py

```
# pip3 install lazypredict
import pandas as pd
from lazypredict.Supervised import LazyClassifier
from sklearn.datasets import load_breast_cancer
from sklearn.model_selection import train_test_split

pd.set_option('display.max_rows', 500)
pd.set_option('display.max_columns', 500)
pd.set_option('display.width', 1000)

df = pd.read_csv("titanic2.csv")
X  = df[["age","class"]]
y  = df[["survived"]]

X_train, X_test, y_train, y_test = train_test_split(X,
y,test_size=.5,random_state =123)

classifier = LazyClassifier(verbose=0,ignore_
warnings=True, custom_metric=None)
models, predictions = classifier.fit(X_train, X_test,
y_train, y_test)
print(models)
```

Listing 4.3 starts with several `import` statements and then sets several Pandas options via the `set_option()` method. The next code block initializes the Pandas data frame `df` with the contents of the CSV file `titanic2.csv`, and then initializes the variable X with the `age` and `class` features of the data frame `df`. In addition, the variable y is initialized with the contents of the survived feature of the data frame `df`.

The next code snippet initializes the arrays X_train, X_test, y_train, x_test, with the result of invoking the `train_test_split()` method in `sklearn.model_selection`. Now we can instantiate the variable classifier as an instance of the `LazyClassifier` class, and then invoke the `fit()` of the classifier instance with the previously defined four arrays.

The final code snippet initializes the variables `models` and `predictions` and then prints the contents of the `models` variable. Launch the code in Listing 4.3, and you will see the following output:

Model	Accuracy	Balanced Accuracy	ROC AUC	F1 Score	Time Taken
NuSVC	0.70	0.68	0.68	0.71	0.03
NearestCentroid	0.65	0.66	0.66	0.66	0.04
Perceptron	0.63	0.65	0.65	0.64	0.04
RandomForestClassifier	0.65	0.61	0.61	0.65	0.20
AdaBoostClassifier	0.70	0.61	0.61	0.68	0.14
KNeighborsClassifier	0.70	0.60	0.60	0.68	0.05
GaussianNB	0.67	0.59	0.59	0.66	0.04
LinearDiscriminantAnalysis	0.73	0.59	0.59	0.68	0.06
LinearSVC	0.73	0.59	0.59	0.68	0.03
SGDClassifier	0.74	0.59	0.59	0.68	0.03
XGBClassifier	0.66	0.59	0.59	0.65	0.09
RidgeClassifier	0.73	0.58	0.58	0.67	0.05
LogisticRegression	0.73	0.58	0.58	0.67	0.05
RidgeClassifierCV	0.74	0.58	0.58	0.67	0.03
LabelSpreading	0.69	0.57	0.57	0.66	0.04
BaggingClassifier	0.64	0.57	0.57	0.64	0.07
LabelPropagation	0.69	0.57	0.57	0.66	0.04
QuadraticDiscriminantAnalysis	0.49	0.57	0.57	0.50	0.06
LGBMClassifier	0.73	0.56	0.56	0.65	0.17
DecisionTreeClassifier	0.59	0.55	0.55	0.60	0.03
PassiveAggressiveClassifier	0.68	0.53	0.53	0.62	0.04
CalibratedClassifierCV	0.70	0.53	0.53	0.61	0.08
ExtraTreesClassifier	0.55	0.53	0.53	0.56	0.14
SVC	0.69	0.51	0.51	0.59	0.03
BernoulliNB	0.69	0.50	0.50	0.57	0.04
DummyClassifier	0.69	0.50	0.50	0.57	0.04
ExtraTreeClassifier	0.53	0.49	0.49	0.54	0.03

What is the TOC Curve?

The following URL contains a `Python` code sample using `SkLearn` and the Iris dataset, and includes code for plotting the ROC:

https://scikit-learn.org/stable/auto_examples/model_selection/plot_roc. html

The following website contains an assortment of *Python code samples for plotting the ROC:*

https://stackoverflow.com/questions/25009284/how-to-plot-roc-curve-in-python

By contrast, a `TOC` graph plots the `(TP+FP)-TP` values on the horizontal axis and the `TP` values for the vertical axis. The interesting fact about a `TOC` graph is that it enables you to determine the confusion matrix for every point in `TOC` space.

ROC AUC and PR AUC

The choice of the ROC AUC versus PR AUC (Precision/Recall AUC) depends on the characteristics of your dataset. In particular, use ROC AUC when both classes are equally important.

However, a highly skewed dataset, or a dataset with a small number of incorrect predictions, can adversely affect both the ROC curve and the ROC AUC curve. In this case, consider using the PR AUC curve, especially when the positive class is more important.

In addition, the PR AUC is well-suited for binary predictions involving imbalanced datasets. The following code snippet shows you how to import `precision_recall_curve` from scikit-learn (note that `ytest` and `yscores` must be initialized for the code to work):

```
from sklearn.metrics import precision_recall_curve
precision, recall,_ = precision_recall_curve(ytest,
scores)
```

THE SKLEARN.METRICS MODULE (OPTIONAL)

If you plan to study machine learning, then the scikit-learn library will probably be one of your most commonly used libraries. This library contains a plethora of metrics for classification algorithms, "under the curve" classes, and a classification report class.

Here are some `import` statements for working with metrics:

- `from sklearn.metrics import accuracy_score`
- `from sklearn.metrics import average_precision_score`
- `from sklearn.metrics import balanced_accuracy_score`
- `from sklearn.metrics import confusion_matrix`
- `from sklearn.metrics import f1_score`
- `from sklearn.metrics import precision_score`
- `from sklearn.metrics import recall_score`

Here are some suggestions for working with curves and reports:

- `from sklearn.metrics import auc`
- `from sklearn.metrics import precision_recall_curve`
- `from sklearn.metrics import roc_auc_score`
- `from sklearn.metrics import roc_curve`
- `from sklearn.metrics import classification_report`

Scikit-learn provides extensive online documentation, along with downloadable code samples that show you how invoke many of the classes that are provided in this library, as well as code samples that calculate the plethora of metrics that are discussed in this chapter.

STATISTICAL METRICS FOR CATEGORICAL DATA

One of the previous sections discussed metrics for continuous data, which appears in linear regression tasks. In a subsequent section, you will learn about metrics for categorical data, such as the F1 score. There are related measures for categorical data, such as F2 score, F3 score, and even F0.5 score, all of which are briefly discussed in the following subsections.

What is an F1 Score?

An F1 score in machine learning is for models that are evaluated on a feature that contains categorical data, whereas the p-value is useful for machine learning in general. An *F1 score* is a measure of the accuracy of a test, and it is defined as the *harmonic mean* of precision and recall. First, let's define the formulas for the precision p and the recall r, followed by the formula for an F1 score (which is based on p and r):

```
p = (# of correct positive results)/(# of all positive
results)
r = (# of correct positive results)/(# of all relevant
samples)

F1-score  = 1/[((1/r) + (1/p))/2]
          = 2*(p*r)/(p+r)
```

Keep in mind that an F1 score is for categorical classification problems, whereas the `R^2` value is typically for regression tasks (such as linear regression). Here is a summary of various aspects of an F1 score:

- F1 is always between p (precision) and r (recall).
- F1 equals the arithmetic mean iff `p = r`.
- F1 is often smaller than the mean of p+r.
- if `p <= r` then `F1 = 2*[p*r]/[p+r] <= 2*r*r/[2*r] = r`
- if `r <= p` then `F1 = 2*[p*r]/[p+r] <= 2*p*p/[2*p] = p`

F2, F3, and Fbeta Scores

The F1 score gives equal weight to the value of the precision and the recall value. Although the formulas for an F2 score and an F3 score are similar to the formula for an F1 score, they assign different "weights" to the precision and recall values. The formulas for the F2 score and F3 score are shown here, along with the differences (with respect to an F1 score) shown in bold:

```
F1-score = 2/[1/r + 1/p]
F2-score = 5/[1/r + 4/p]
F3-score = 10/[9/r + 1/p]
```

Finally, there is the *Fbeta measure*, which is defined in terms of p, r, and a third parameter called `beta`, as shown here:

```
Fbeta = ((1 + beta**2)*p*r)/(beta**2*p+r)
```

In particular, when we set `beta` equal to 0.5, we obtain the F0.5 measure, as shown here:

```
F0.5 = ((1 + 1/4)*p*r)/(1/4*p + r)
F0.5 = ((5/4)*p*r)/((p + 4*r)/4)
F0.5 = (5*p*r)/(p + 4*r)
```

Compare the denominator (p+r) with (p+4*r), and you can see that the latter term decreases because r is multiplied by 4, whereas the former term equally balanced for p and r. Hence, the F0.5 score is weighted more toward precision, and less so for the recall.

Listing 4.5 displays the content of the Python file F1Beta.py that shows you how to calculate an F1 beta score.

Listing 4.5: F1Beta.py

```
from sklearn.metrics import fbeta_score


y_true = [1, 1, 0, 1, 1, 0, 1, 1]
y_pred = [0, 1, 0, 0, 1, 0, 1, 0]
result = fbeta_score(y_true, y_pred,beta=0.5)
print("F1Beta score:",result)
```

Listing 4.5 contains an import statement followed by the initialization of the variables y_pred and y_true as two lists of values of equal length. Next, the variable result is initialized with the result of invoking the method fbeta_score() with y_pred and y_true. Launch the code in Listing 4.5, and you will see the following output:

```
F1Beta score: 0.8333333333333334
```

Listing 4.6 displays the content of the Python file F1Score.py that shows you how to calculate an F1 score.

Listing 4.7: F1Score.py

```
from sklearn.metrics import f1_score


y_true = [1, 1, 0, 1, 1, 0, 1, 1]
y_pred = [0, 1, 0, 0, 1, 0, 1, 0]


result = f1_score(y_true, y_pred)
print("F1 score:",result)
```

Listing 4.7 contains an import statement followed by the initialization of the variables y_pred and y_true as two lists of values of equal length. Next, the variable result is initialized with the result of invoking the method `f1_score()` with `y_pred` and `y_true`. Launch the code in Listing 4.7, and you will see the following output:

```
F1 score: 0.6666666666666666
```

As you can see, the F1 beta score is significantly higher than the corresponding F1 score displayed in the output from the previous section.

Guidelines for F1-related Scores

Here are some guidelines that will help you decide which F1-based metric to use, where FP and FN represent false positive and false negative, respectively:

- If FP and FN are equally costly, use an F1 score.
- If FP is more costly, use an F0.5 score.
- If FN is more costly, use an F2 score.

Brier Score

The *Brier Score* measures the accuracy of probabilistic predictions, which is the same as the mean squared error in the univariate case. The formula for the Brier score is shown here:

```
Brier Score = (f_prob - outcome)**2 where:
f_prob = forecasted probability
outcome = 1 if the event occurs, 0 otherwise
```

The Brier Score is designed for scenarios in which predictions involve probabilities that are made on mutually and exclusive discrete outcomes. As you probably expect, the set of assigned probabilities must conform to a discrete probability distribution. Moreover, a lower Brier Score indicates a better set of predictions.

Listing 4.8 displays the content of the Python file `brier1.py` that shows you how to calculate a Brier Score.

Listing 4.8: brier1.py

```
import numpy as np

y_vals  = np.array([1, 0, 1, 1, 1, 0, 0, 1, 1, 1])
y_preds = np.array([0.31, 0.22, 0.83, 0.74, 0.91, 0.23,
0.56, 0.76, 0.73, 0.97])
losses  = np.subtract(y_vals, y_preds)**2
```

```
brier_score = losses.sum()/len(y_vals)
print("brier score:",brier_score)
print("losses:      ",losses)
```

Listing 4.8 contains an `import` statement followed by the initialization of the variables `y_vals` and `y_preds` as two NumPy arrays of equal length. Next, the variable `losses` is initialized with the result of invoking the NumPy method `subtract()` with `y_pred` and `y_true`, and then the program squares the result.

The next portion of Listing 4.8 initializes the variable `brier_score` with the sum of the values in the variable losses, divided by the length of the NumPy array `y_vals`. Launch the code in Listing 4.8, and you will see the following output:

```
brier score: 0.11269999999999998
losses:      [0.4761 0.0484 0.0289 0.0676 0.0081 0.0529
0.3136 0.0576 0.0729 0.0009]
```

METRICS FOR CONTINUOUS DATA

The statistical quantities in this section pertain to continuous data. In particular, this section describes terms that are used in linear regression. A subsequent section discusses statistical terms, such as an F1 score, that pertain to categorical data.

RSS, TSS, and R^2

The following list of items contains statistical terms that are calculated for continuous data (such as linear regression):

- RSS
- TSS
- R^2

The term RSS is the "residual sum of squares" and the term TSS is the "total sum of squares." These terms are used in regression models.

As a starting point so we can simplify the explanation of the preceding terms, suppose that we have a set of points `{(x1,y1), . . . , (xn,yn)}` in the Euclidean plane. In addition, let's define the following quantities:

- `(x,y)` is any point in the dataset.

- y is the y-coordinate of a point in the dataset.
- y_ is the mean of the y-values of the points in the dataset.
- y_hat is the y-coordinate of a point on a best-fitting line.

Just to be clear, (x,y) is a point in the *dataset*, whereas (x,y_hat) is the corresponding point that lies on the *best fitting line*. With these definitions in mind, the definitions of RSS, TSS, and R^2 are listed here (n equals the number of points in the dataset):

$$RSS = Iy - \textbf{y\_hat})**2/n \quad \frac{\Sigma\left(y - y_{hat}\right)\,\char94\,2}{n}$$

$$TSS = (y - \textbf{y\_bar})**2/n \quad \frac{\Sigma\left(y - y_{bar}\right)\,\char94\,2}{n}$$
$$R\char94 2 = 1 - RSS/TSS$$

We also have the following inequalities involving RSS, TSS, and R^2:

```
0 <= RSS
RSS <= TSS
0 <= RSS/TSS <= 1
0 <= 1 - RSS/TSS <= 1
0 <= R^2 <= 1
```

When RSS is close to 0, then RSS/TSS is also close to zero, which means that R^2 (R^2) is close to 1. Conversely, when RSS is close to TSS, then RSS/TSS is close to 1, and R^2 is close to 0. In general, a larger R^2 is preferred (i.e., the model is closer to the data points), but a lower value of R^2 is not necessarily a bad score.

Adjusted R^2

An *adjusted* R^2 (R^2) value, let's call it ARS, is calculated by a formula that is based on R^2, n (the number of samples), and k (the number of independent variables), as shown here:

$$ARS = 1 - \frac{(1-R\char94 2)*(n-1)}{(n-k-1)}$$

In the case of linear regression in the Euclidean plane, ARS = R^2 because k = 1, so the preceding formula for ARS becomes ARS = 1 - (1-R^2) = R^2.

If k > 1, set z = (n-1)/(n-k-1), and since n-k-1 < n-1, then z > 1, and the formula for ARS becomes the following:

```
ARS = 1 - z*(1-R^2) = 1 - z + z*R^2 > z*R^2 > R^2.
```

Consequently, ARS > R^2 whenever k > 1.
The following results are also true:

1. `ARS > 0 if R^2 < k/(N-1)`

2. `ARS = 0 if R^2 = k/(N-1)`

3. `ARS < 0 if R^2 > k/(N-1)`

Let's derive the third result:

```
if R^2 > k/(N-1), then 1-R^2 < 1-k/(N-1) = [(N-1)-k]/
(N-1), and so we get:
```

```
             (N-1)
ARS = 1 - ------- * (1-R^2)
           (N-1-k)


            (N-1)
   < 1 - ------- * (1-k/(N-1))
          (N-1-k)


            (N-1)
   = 1 - ------- * [(N-1-k)/(N-1)] = 1 - 1 = 0 QED
          (N-1-k)
```

R^2 and Its Limitations

One of the most frequently used metrics is R^2 ("R-squared"), which measures how close the data is to the fitted regression line (regression coefficient). The R^2 value is always a percentage between 0 and 100%. The value 0% indicates that the model explains none of the variability of the response data around its mean. The value 100% indicates that the model explains all the variability of the response data around its mean. In general, a higher R^2 value indicates a better model.

Although high R^2 values are preferred, they are not necessarily always good values. Similarly, low R^2 values are not always bad. For example, an R^2 value for predicting human behavior is often less than 50%. Moreover, R^2 cannot determine whether the coefficient estimates and predictions are biased. In addition, an R^2 value does not indicate whether a regression model is adequate. Thus, it is possible to have a low R^2 value for a good model, or a high R^2 value for a poorly fitting model. Evaluate R^2 values in conjunction with residual plots, other model statistics, and subject area knowledge.

MAE, MSE, AND RMSE

The following list expands the initialisms MAE, MSE, and RMSE:

- MAE is an initialism for the Mean Absolute Error.
- MSE is an initialism for the Mean Squared Error.
- RMSE is an initialism for the Root Mean Squared Error.

In addition, the term RSS is an initialism for "Residual Sum of Squares," and the term TSS is an initialism for the "Total Sum of Squares." The possible values of R^2 lie between 0 and 1, with higher values associated with better models.

In simple terms, the MSE is the sum of the squares of the difference between *actual* y values and the *predicted* y values, divided by the number of points. Notice that the predicted y value is the y value that each point would have if that point were actually on the best-fitting line.

The MSE is the basis for the preceding error types. For example, RMSE is the "Root Mean Squared Error," which is the square root of the MSE. The value of the MSE is the sum of the variance and the square of the bias of a dataset.

The MAE is the "Mean Absolute Error," which is the sum of *the absolute value of the differences of the y terms* (*not* the square of the differences of the y terms), which is then divided by the number of terms.

Although it is easier to compute the derivative of the MSE, it is also true that the MSE is more susceptible to outliers, whereas the MAE is less susceptible to outliers. The reason is simple: a squared term can be significantly larger than the absolute value of a term. For example, if a difference term is 10, then a squared term of 100 is added to the MSE, whereas only 10 is added to the MAE. Similarly, if a difference term is -20, then a squared term 400 is added to the MSE, whereas only 20 (which is the absolute value of -20) is added to the MAE.

The following article provides a more detailed explanation of the MAE, MSE, and RMSE:

https://www.datatechnotes.com/2019/02/regression-model-accuracy-mae-mse-rmse.html

A Code Sample for the MAE, MSE, and RMSE

Listing 4.4 displays the content of the Python file `mae_mse_rmse.py` that shows you how to calculate the MAE, MSE, and RMSE of a set of numbers.

Listing 4.4: mae_mse_rmse.py

```
from sklearn.metrics import mean_absolute_error as mae
values  = [12, 13, 14, 15, 15, 22, 27]
predict = [11, 13, 14, 14, 15, 16, 18]

# 1) mean absolute error:
mae1 = mae(values, predict)
print("mae1:",mae1)

import numpy as np
Y_true = [1,1,2,2,4]  # Y_true = Y (original values)
Y_pred = [0.6,1.29,1.99,2.69,3.4]

# 2) Mean Squared Error:
mse1 = np.square(np.subtract(Y_true,Y_pred)).mean()
print("mse1:",mse1)

# 3) root mean squared error:
from sklearn.metrics import mean_squared_error as mse
import math
values  = [0, 1, 2, 0, 3]
predict = [0.1, 1.3, 2.1, 0.5, 3.1]
```

```
mse1 = mse(values, predict)
rmse = math.sqrt(mse1)
print("rmse:",rmse)
print()
```

Listing 4.4 starts with an `import` statement, followed by four blocks of code that calculate and then print the values of the MAE, MSE, and RMSE. Launch the code in Listing 4.4, and you will see the following output:

```
mae1: 2.4285714285714284
mse1: 0.21606
rmse: 0.2720294101747089
```

Non-Linear Least Squares

When predicting housing prices where the dataset contains a wide range of values, techniques such as linear regression or random forest can cause the model to overfit the samples with the highest values to reduce quantities such as the MAE.

In this scenario, you probably want an error metric, such as relative error, that reduces the importance of fitting the samples with the largest values. This technique is called *non-linear least squares*, which may use a log-based transformation of labels and predicted values.

APPROXIMATING LINEAR DATA WITH NP.LINSPACE()

Listing 4.5 displays the content of `np_linspace1.py` that illustrates how to generate some data with the `np.linspace()` API in conjunction with the "perturbation technique."

Listing 4.5: np_linspace1.py

```
import numpy as np

trainX = np.linspace(-1, 1, 6)
trainY = 3*trainX + np.random.randn(*trainX.shape)*0.5

print("trainX: ", trainX)
print("trainY: ", trainY)
```

The purpose of this code sample is merely to generate and display a set of randomly generated numbers. (This code can be used as a starting point for an actual linear regression task.)

Listing 4.5 starts with the definition of the array variable `trainX` that is initialized via the `np.linspace()` API. Next, the array variable `trainY` is defined via the "perturbation technique" that you saw in previous code samples. The output from Listing 4.5 is here:

```
trainX:  [-1.  -0.6 -0.2  0.2  0.6  1. ]
trainY:  [-2.9008553  -2.26684745 -0.59516253
0.66452207  1.82669051  2.30549295]
trainX:  [-1.  -0.6 -0.2  0.2  0.6  1. ]
trainY:  [-2.9008553  -2.26684745 -0.59516253
0.66452207  1.82669051  2.30549295]
```

SUMMARY

This chapter started an introduction to the ROC curve and AUC curve. You also learned about metrics for linear regression, such as the MAE, MSE, RMSE, and MMR.

Next, you learned about type I errors, type II errors, and the metrics accuracy, precision, and recall for categorical data. Finally, you learned about the F1 score and related scores, such as F2, and F3, which are useful for classification algorithms in machine learning.

## CHAPTER 5

PROBABILITY DISTRIBUTIONS

This chapter provides an overview of some well-known discrete probability distributions as well as continuous probability distributions. You will see Python code samples for various distributions, along with a generated image with a sample output.

The first section of this chapter starts with an explanation of the PDF, CDF, and PMF, followed by an introduction to discrete probability distributions, such as the Bernoulli, binomial, and Poisson distributions.

The second section introduces continuous probability distributions, such as the chi-squared, Gaussian, and uniform distributions. This section also discusses non-Gaussian distributions and some of the causes of such distributions.

In case you have not already done so, please make sure that the NumPy, Matplotlib, and SciPy libraries are installed in your Python environment by launching the following commands from the command line:

```
pip3 install numpy matplotlib scipy
```

Most of the code samples for generating probability distributions in this chapter were generated via GPT4 (unless indicated otherwise).

PDF, CDF, AND PMF

PDF, CDF, and PMF are initialisms for two probability functions and a cumulative distribution function that are important in statistics.

- PDF (probability density function)
- CDF (cumulative distribution function)
- PMF (probability mass function)

A *probability density function* (PDF) is a statistical function used in continuous probability distributions to describe the likelihood of a random variable taking on a particular value. Unlike discrete probability distributions, where probabilities are defined for distinct, individual outcomes, continuous distributions involve a range of outcomes. A PDF has nonnegative values and the area under a PDF equals 1. In addition, a CDF can be expressed as an integral of the PDF in the continuous case.

A PMF is a function that gives the probability of each possible outcome for a discrete random variable. Recall that unlike continuous random variables, which have a PDF, discrete random variables have distinct, individual outcomes, each with a specific probability of occurring. PDFs and CDFs are alternatives to binning data in datasets as a way to avoid binning bias.

For example, given a random variable X that has a Gaussian distribution as its PDF, then `P(X<=mean)` is the probability that the random variable X is less than or equal to the mean. The Gaussian distribution is symmetric, and therefore the mean and the median are equal, which in turn implies that `P(X<=mean)= 0.5`.

A *cumulative density function* (CDF) is a function, let's call it `F(x)`, that has the form `F(x) = P(X ≤x)`, which provides the probability that the random variable X <= x, for a given value of x. For example, the CDF for tossing a balanced die is as follows:

```
P(X<1)  = 0
P(X<=1) = 1/6
P(X<=2) = 2/6
P(X<=3) = 3/6
P(X<=4) = 4/6
P(X<=5) = 5/6
P(X<=6) = 6/6
P(X>6)  = 0
```

A *probability mass function* (PMF) applies to the discrete domain and gives the probability that a discrete random variable is exactly equal to some value.

In addition, the PMF is a probability distribution, which means that all probabilities are nonnegative (zero is permitted), and the sum of the probabilities of all elements in the sample space equals 1. Simple examples of PMFs include

- tossing a fair coin (two probabilities that equal 1/2)
- throwing a balanced die (six probabilities that equal 1/6)

For example, the following set contains the discrete probabilities of tossing two dice that can yield the values {2, 3, 4, . . . , 12}:

{1/36,2/36,3/36,4/36,5/36,6/36,5/36,4/36,3/26,2/36,1/36}

TWO TYPES OF PROBABILITY DISTRIBUTIONS

In statistics, there are two main types of probability distributions: discrete and continuous. The majority of the remaining sections in this chapter describe both types of probability distributions. Familiarity with both types of distributions can assist you in identifying the distribution that most closely matches the data in a dataset, as discussed in the next section.

Selecting a Distribution

Datasets can vary from small (two columns with 100 rows) to extremely large (10,000 columns with one billion rows). Thus, selecting a distribution that matches a dataset can be simple for small datasets and challenging for larger datasets. In general, here are several steps that you can follow to select a distribution for a dataset:

- Plot the data in a histogram and make inferences.
- Make assumptions about the distribution of the data.
- Find the closest matching distribution from a list of well-known distributions.

This process often involves a combination of trial-and-error and guesswork, potentially involving multiple attempts to find a suitable distribution. Keep in mind that the distribution you select might only be a modest (close-enough?) approximation to the distribution of the data in the dataset.

There might be multiple distributions that are reasonable approximations to a given dataset. In this case, you can perform additional tests to determine the best-fitting distribution for a given dataset.

Another approach that could significantly simplify your task involves selecting a small number of relevant columns in a dataset, which is well-suited for datasets with a large number of columns. There are various techniques in machine learning for selecting a subset of columns that are the most significant for a given dataset. Such techniques are outside the scope of this book, but you can find online articles that provide useful information.

Now let's turn our attention to some well-known discrete probability distributions, as discussed in the next section.

DISCRETE PROBABILITY DISTRIBUTIONS

The following list contains some well-known discrete probability distributions:

- Bernoulli
- binomial
- Poisson

The following subsections contain more information about the distributions in the preceding list.

The Bernoulli Distribution

A *Bernoulli distribution* involves a random variable X with 2 possible outcomes: 1 (success) occurs with probability p and 0 (failure) occurs with probability 1-p. Here are some relevant formulas for a single event:

```
P(X=1) = p
P(X=0) = 1-p
E[X]   = p
VAR[X] = p*(1-p)
```

In the case of n Bernoulli variables, we have the following formulas:

```
E[X]   = n*p
VAR[X] = n*p*(1-p)
```

Notice that when p = 1/2, we have a distribution for a fair coin.

The value of E[X] can be inferred from another example. Suppose you toss a fair coin 100 times, and you receive $1 for each head and $0 for each tail. Since you win money only half the time, E[X] is as follows:

```
E[X] = 100 * [1 * (1/2) + 0 * (1/2)] =  100 * (1/2) = 50
```

Replacing 100 with n gives the following result:

```
E[X] = n * [1 * (1/2) + 0 * (1/2)] =  n * (1/2) = n*p
```

Figure 5.1 shows the graph of the Bernoulli distribution.

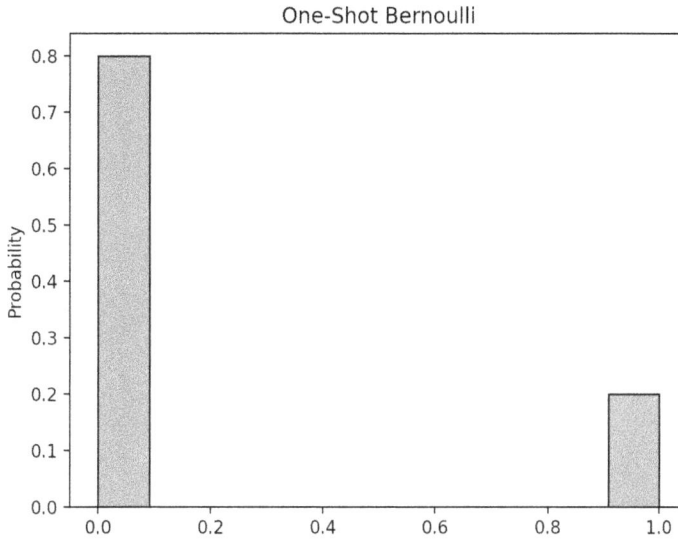

FIGURE 5.1: An Example of a Bernoulli distribution

The Binomial Distribution

A *binomial distribution* is the probability distribution of the number of successes in a sequence of n independent experiments, each of which can have only 2 possible outcomes, as described in a Bernoulli distribution.

An example of a binomial distribution involves the number of heads tossed in n coin tosses (a single coin toss is a Bernoulli distribution). There are only two possible outcomes for each coin toss, both of which can occur with a probability of 0.5. However, a coin toss is just one example of a binomial distribution: the general case involves a random variable X whose probability of occurrence is p, where p can be any value between 0 and 1 (i.e., not just 0.5).

A binomial distribution involves multiple Bernoulli experiments, with the parameter p (the probability of success) and the parameter n (the number of repetitions). In addition, the binomial distribution is left-skewed, symmetric, and right-skewed when p is greater than 0.5, equal to 0.5, and less than 0.5, respectively.

Let's display a binomial distribution to show that it is similar to a discrete approximation to a Gaussian distribution for p = 0.5 and for large values of n. In fact, the binomial distribution approximates the area under the curve as n becomes arbitrarily large.

Figure 5.2 shows the graph of the binomial distribution.

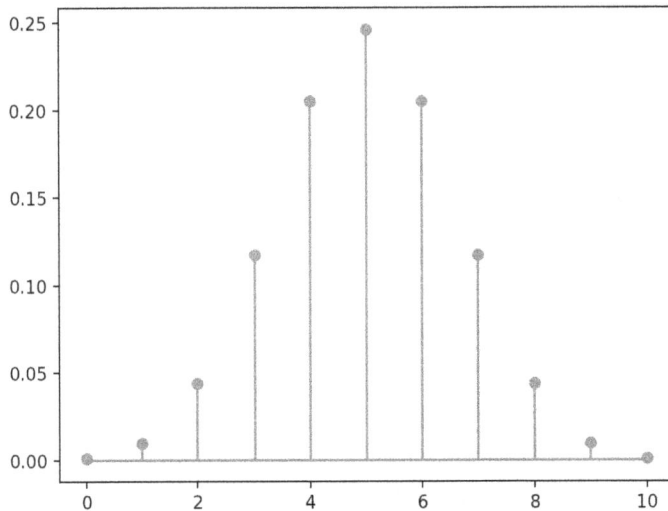

FIGURE 5.2: A binomial distribution

Sample Values of a Binomial Distribution

This section contains a simple scenario for calculating cumulative probabilities involving coin tosses. Specifically, suppose that you toss a balanced coin n times, where the outcome of each coin toss is heads with a probability of 0.5 and tails with a probability of 0.5. In addition, suppose that you want to know the probability of displaying, at most, k heads during the n coin tosses, where n = 4. Thus, we want to determine the cumulative probability of the following:

- at most 0 heads in 4 coin tosses
- at most 1 head in 4 coin tosses
- at most 2 heads in 4 coin tosses
- at most 3 heads in 4 coin tosses
- at most 4 heads in 4 coin tosses

The calculation of the preceding values involves combinatorial coefficients, as shown here for n=4:

- C(4,0) ways to generate a sequence of 0 heads in 4 coin tosses: 1
- C(4,1) ways to generate a sequence of 1 head in 4 coin tosses: 4

- C(4,2) ways to generate a sequence of 2 heads in 4 coin tosses: 6
- C(4,3) ways to generate a sequence of 3 heads in 4 coin tosses: 4
- C(4,4) ways to generate a sequence of 4 heads in 4 coin tosses: 1

Thus, the combinatorial coefficients C(4,k) for `k` in `{0,1,2,3,4}` are 1 4 6 4 1. Next, the cumulative probabilities are calculated below, where `n` equals the number of coin tosses between 0 and 4, inclusive (the left-side values are approximations), that are calculated for the first two values of `n`:

```
P(X <= 0) = (0.5)^4 = 0.0625
```

```
P(X <= 1) = C(4,2)*0.062 + predecessor probability
          = 6*0.062 + 1*0.312
          = 0.372 + 0.312 = 0.684
```

If you are wondering why the probability of 0 heads equals 0.0625 for 4 coin tosses, this number equals the probability of 4 tails, which is calculated as follows:

```
(1/2)*(1/2)*(1/2)*(1/2) = 1/16 = 0.0625.
```

Values of a Binomial Distribution for Different Values of p

The preceding section showed you how to calculate the cumulative binomial values for `n = 4` and `p = 0.5`. Specifically, the probability of tossing 0 heads is reproduced here:

```
n = 0: P(X = 0) = (0.5)^4 = 0.0625
```

If you want to use a probability value of 1/4, then you need to replace 0.5 with 0.25 in the preceding code snippet, which results in the following:

```
n = 0: P(X = 0) = (0.25)^4 = 1/256 = 0.0039
```

The Poisson Distribution

A *Poisson distribution* is a discrete probability distribution that provides the probability of a given number of events `k` occurring in a fixed interval of time. In addition, these events occur with a known constant average rate λ and independently of the time.

The *Poisson distribution* is a skewed probability distribution that has real life applications, such as counting the number of cars that travel past a selected point on the side of a road during a given period of time. A second example involves the number of people who go to a public pool based on five-degree

(Fahrenheit) increments of the temperature, which is a set of numbers that follows a Poisson distribution.

The formula for a Poisson distribution involves two independent parameters λ and k, as shown here:

```
P(x) = λ^x*exp(-k*x)/x!
```

In the preceding formula, the parameter λ is called the "event rate" and k specifies the number of occurrences. Interestingly, the mean and variance of the Poisson distribution are as follows:

```
mean = variance = λ
```

Generating Poisson Values

As an example, Listing 5.1 shows you the content of `poisson1.py` that generates values from a Poisson distribution.

Listing 5.1: poisson1.py

```
from scipy import stats


for kval in range (3,6):
  for muval in range(0,3):
     print("kval: ",kval," mu:", muval, "stats.poisson.
pmf:", stats.poisson.pmf(k=kval, mu=muval))
     print("----------------\n")
```

Listing 5.1 starts with an `import` statement followed by a loop that iterates through the integers from 3 to 5, inclusive. During each iteration, the code executes another loop that displays a set of values for mu (from 0 to 2, inclusive) and the corresponding pmf value. Launch the code in Listing 5.1, and you will see the following output:

```
kval: 3 mu: 0 stats.poisson.pmf: 0.0
kval: 3 mu: 1 stats.poisson.pmf: 0.06131324019524039
kval: 3 mu: 2 stats.poisson.pmf: 0.18044704431548356

kval: 4 mu: 0 stats.poisson.pmf: 0.0
kval: 4 mu: 1 stats.poisson.pmf: 0.015328310048810101
kval: 4 mu: 2 stats.poisson.pmf: 0.09022352215774178
```

```
kval: 5 mu: 0 stats.poisson.pmf: 0.0
kval: 5 mu: 1 stats.poisson.pmf: 0.00306566200976202
kval: 5 mu: 2 stats.poisson.pmf: 0.03608940886309672
-----------------
```

A table of cumulative Poisson distribution values is accessible online:

https://eduschool40.blog/wp-content/uploads/2018/03/Poisson-Table_0.pdf

Rendering a Poisson Distribution

Listing 5.2 shows you a Python code sample that generates a graph of a Poisson distribution.

Listing 5.2: poisson_distribution.py

```python
import numpy as np
import matplotlib.pyplot as plt
from scipy.stats import poisson

# Poisson parameter
lambda_param = 10

# Generate a Poisson distribution
x = np.arange(poisson.ppf(0.01, lambda_param),
              poisson.ppf(0.99, lambda_param))
pmf = poisson.pmf(x, lambda_param)

# Create a bar chart
plt.bar(x, pmf)
plt.title('Poisson Distribution (λ = 10)')
plt.xlabel('Number of Events')
plt.ylabel('Probability')

# Show the plot
plt.show()
```

Listing 5.2 contains several import statements and then initializes the scalar variable `lambda_param` with the value 10, which represents the average rate of success (also known as the *rate parameter*). The success rate pertains to the type of entity that you want to observe, such as the number of cars that pass a given point on an hourly basis or the average number of new users that register on a website on a daily basis.

The next portion of Listing 5.2 uses the np.arange() method in NumPy to generate a range of numbers. In addition, the `poisson.pmf` function from the `scipy.stats.poisson` module is relevant for the range of numbers on the x-axis. This function takes a percentile and a lambda parameter to calculate the PMF for the given values. Launch the code in Listing 5.2, and you will see the graph shown in Figure 5.3.

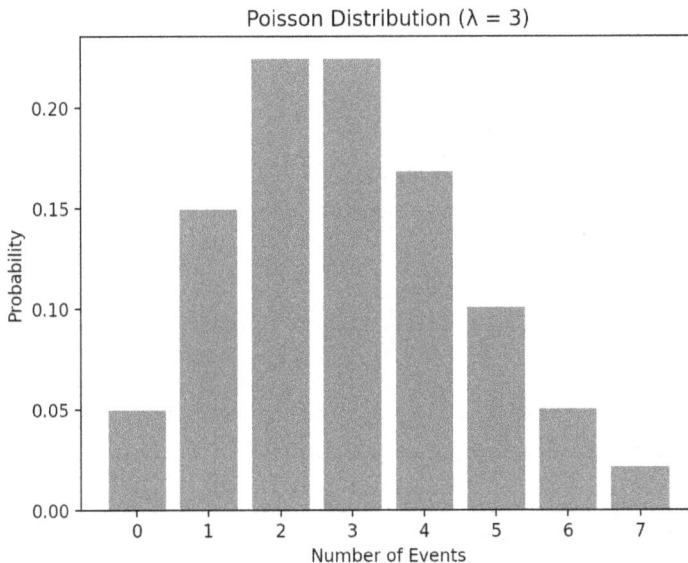

FIGURE 5.3: A Poisson distribution

The Geometric Distribution

A *geometric distribution* is a discrete probability distribution of a random variable X that has two possible outcomes (e.g., heads or tails from flipping a fair coin). The probability of the two outcomes p, and (1-p) is the same for all trials, where p is a positive number between 0 and 1. Moreover, the probability of success occurs during the kth trial is specified by P(X = k) = (1-p)**(k-1)*p.

As a simple example, you can use a geometric distribution to model the number of attempts needed to make a sale.

Listing 5.3: geometric.py

```python
import numpy as np
import matplotlib.pyplot as plt
from scipy.stats import geom

# Geometric distribution parameter
p = 0.5  # probability of success

# Generate a range of x values
x = np.arange(1, 11)

# Generate the corresponding y values from the
geometric distribution
pmf = geom.pmf(x, p)

# Create a bar chart
plt.bar(x, pmf)
plt.title('Geometric Distribution (p=0.5)')
plt.xlabel('Number of Trials')
plt.ylabel('Probability')

# Show the plot
plt.show()
```

Listing 5.3 starts with `import` statements and then initializes the scalar variable `p`, which represents the probability of success on each trial for the geometric distribution. The `geom.pmf` function calculates the probability mass function of the geometric distribution for the given values. Launch the code in Listing 5.3, and you will see the geometric distribution displayed in Figure 5.4.

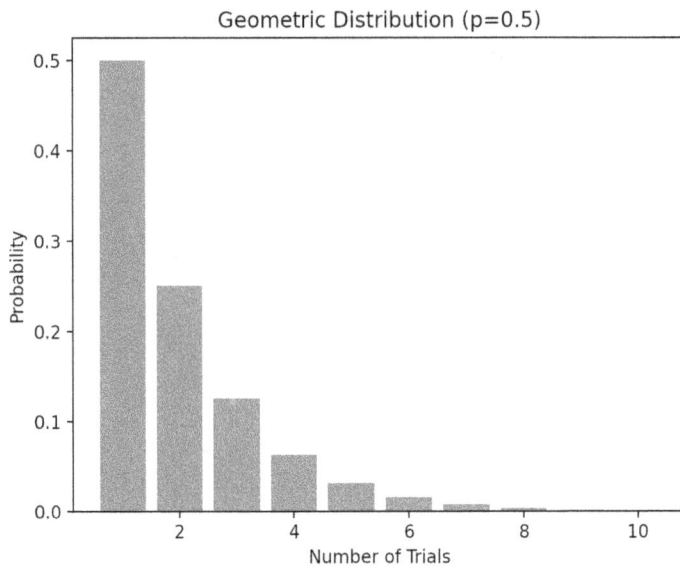

FIGURE 5.4: A geometric distribution

This concludes the portion of the chapter devoted to discrete probability distributions. Now let's turn our attention to some well-known continuous probability distributions, as discussed in the next section.

CONTINUOUS PROBABILITY DISTRIBUTIONS

The following list contains some well-known continuous probability distributions, some of which are discussed in subsequent sections:

- chi-squared
- exponential
- F-distribution
- gamma
- Gaussian
- lognormal
- t-distribution
- uniform

Chi-squared Distribution

A *chi-squared* goodness of fit test determines whether sample data matches a population. In addition, a chi-squared test for independence compares two variables in a contingency table to find if they are related.

Listing 5.4: chisquared_distribution.py

```
import numpy as np
import matplotlib.pyplot as plt
from scipy.stats import chi2

# Chi-squared distribution parameter
df = 4  # degrees of freedom

# Generate a range of x values
x = np.linspace(chi2.ppf(0.01, df), chi2.ppf(0.99, df),
100)

# Generate the corresponding y values from the chi-
squared distribution
pdf = chi2.pdf(x, df)

# Plot the distribution
plt.plot(x, pdf, label='chi2 pdf')
plt.title('Chi-squared Distribution (df=4)')
plt.xlabel('Value')
plt.ylabel('Probability Density')

# Show the plot
plt.show()
```

Listing 5.4 starts with several `import` statements, followed by the scalar variable `df` that represents the degrees of freedom parameter for the chi-squared distribution. The range of x values is determined using a combination of the NumPy `linspace()` method and the percent point function `chi2.ppf`, which is the inverse of the CDF.

The `chi2.ppf` function calculates the PDF of the chi-squared distribution for those x values. Launch the code in Listing 5.4, and you will see the chi-squared graph shown in Figure 5.5.

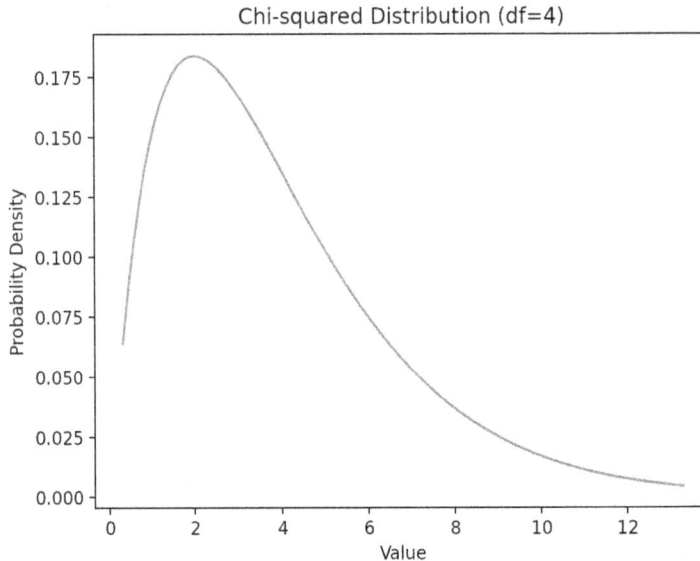

FIGURE 5.5: A chi-squared distribution

The Exponential Distribution

The exponential distribution has ties to the Poisson distribution, the geometric distribution, and the gamma distribution. First, the exponential distribution is the probability distribution of time between consecutive events in the Poisson point process. Second, the exponential distribution is a continuous counterpart of the geometric distribution. Third, the exponential distribution is also a special case of the gamma distribution. Moreover, the formula for an exponential distribution bears a slight resemblance to a Poisson distribution, as shown here:

```
f(x) = lmd * exp(-lmd*x) for x >= 0 and 0 for x < 0
```

Note that `lmd` ("lambda") is a positive value.

An Exponential Distribution Code Sample

Listing 5.5 shows you an example of calculating values from an exponential distribution and then displaying an associated graph.

Listing 5.5: exponential_distribution.py

```
import numpy as np
import matplotlib.pyplot as plt
from scipy.stats import expon


# Exponential distribution parameter
scale = 1.0  # This is 1/lambda, lambda being the rate
parameter


# Generate a range of x values
x = np.linspace(expon.ppf(0.01, scale=scale),
                expon.ppf(0.99, scale=scale),
                100)


# Generate the corresponding y values from the
Exponential distribution
pdf = expon.pdf(x, scale=scale)


# Plot the distribution
plt.plot(x, pdf, label='expon pdf')
plt.title('Exponential Distribution (λ=1.0)')
plt.xlabel('Value')
plt.ylabel('Probability Density')


# Show the plot
plt.show()
```

Listing 5.5 starts with several `import` statements, followed by the scalar variable `scale` that represents the standard deviation of the distribution. In the case of an exponential distribution, `sigma = 1/lambda` (λ). The range of x values is determined using the percent point function `ppf`, which is the inverse of the CDF. The `expon.pdf` function calculates the PDF of the exponential distribution for those x values. Launch the code in Listing 5.5, and you will see the exponential distribution shown in Figure 5.6.

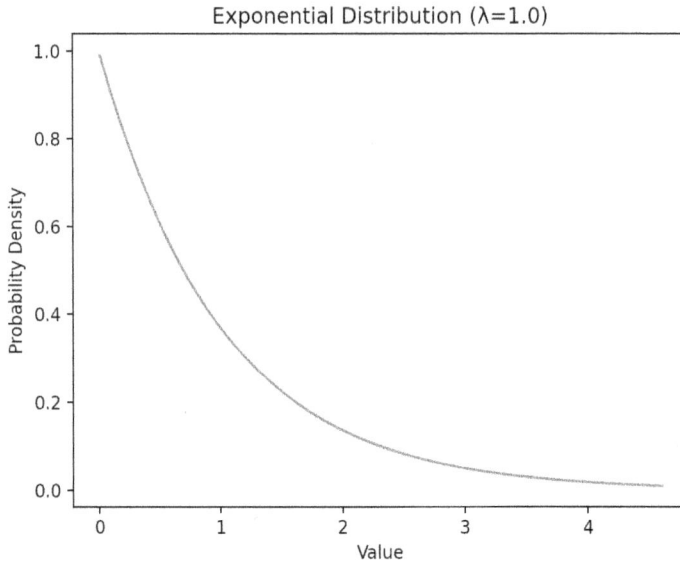

FIGURE 5.6: An exponential distribution

The F-Distribution

An *F-distribution* is defined as a ratio of two chi-squared distributed random variables `chi1` and `chi2` divided by their corresponding degrees of freedom. Two random variables `S1` and `S2` have a chi-squared distribution with `d1` and `d2` degrees of freedom, respectively. Then the random variable `X` defined here has an F-distribution:

```
        S1/d1
X  =  ———————-
        S2/d2
```

Listing 5.6 displays the content of `f_distribution.py` that shows you how to render one F-distribution.

Listing 5.6: f_distribution.py

```python
import numpy as np
import matplotlib.pyplot as plt
from scipy.stats import f


# F distribution parameters
dfn = 15  # numerator degrees of freedom
dfd = 15  # denominator degrees of freedom


# Generate a range of x values
x = np.linspace(f.ppf(0.01, dfn, dfd), f.ppf(0.99, dfn,
dfd), 100)


# Generate the corresponding y values from the F
distribution
pdf = f.pdf(x, dfn, dfd)


# Plot the distribution
plt.plot(x, pdf, label='F pdf')
plt.title('F Distribution (dfn=15, dfd=15)')
plt.xlabel('Value')
plt.ylabel('Probability Density')


# Show the plot
plt.show()
```

Listing 5.6 starts with `import` statements followed by the variables `xdfn` and `dfd` that represent degrees of freedom. Next, array `x` is initialized as a range of values by a combination of the NumPy `linspace()` method and the `ppf()` method.

The next portion of Listing 5.6 initializes the variable `pdf` via the `pdf()` method and then plots the graph with a title and two labeled axes. Launch the code in Listing 5.6, and you will see the F-distribution shown in Figure 5.7.

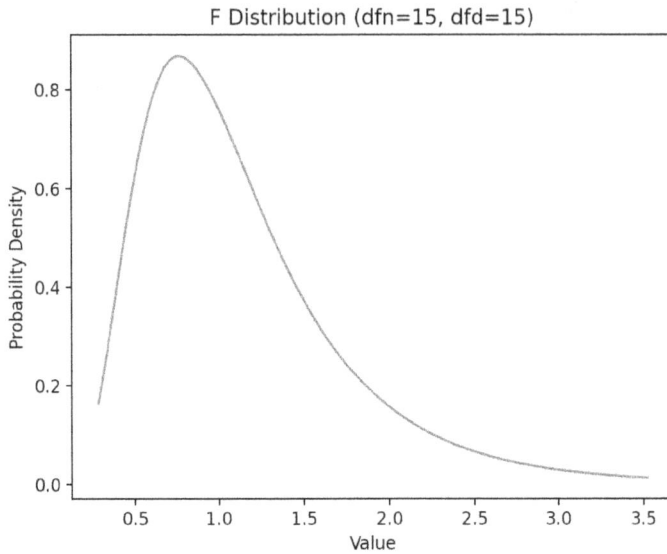

FIGURE 5.7: An F-distribution

The Gamma Distribution

A *gamma distribution* is related to the beta distribution. A gamma distribution is a widely used distribution that can model continuous variables. A gamma distribution is a sophisticated distribution whose parameter values can generate a rich set of functions. More information about a gamma distribution (mean, variance, and so forth) is accessible online:

https://byjus.com/maths/gamma-distribution

Listing 5.7: gamma.py

```
import numpy as np
import matplotlib.pyplot as plt
from scipy.stats import gamma

# Gamma distribution parameters
shape = 2.0  # shape parameter, often known as k
scale = 2.0  # scale parameter, often known as theta

# Generate a range of x values
x = np.linspace(gamma.ppf(0.01, shape, scale=scale),
```

```
        gamma.ppf(0.99, shape, scale=scale),
        100)


# Generate the corresponding y values from the Gamma
distribution
pdf = gamma.pdf(x, shape, scale=scale)


# Plot the distribution
plt.plot(x, pdf, label='gamma pdf')
plt.title('Gamma Distribution (k=2.0, θ=2.0)')
plt.xlabel('Value')
plt.ylabel('Probability Density')


# Show the plot
plt.show()
```

Listing 5.7 starts with `import` statements followed by the variables' shape and scale that represent the shape (k) and scale (θ) parameters for the gamma distribution. The range of x values is determined using the percent point function `ppf`, which is the inverse of the CDF. The `gamma.pdf` function calculates the PDF of the gamma distribution for those x values. Launch the code in Listing 5.7, and you will see the gamma distribution shown in Figure 5.8.

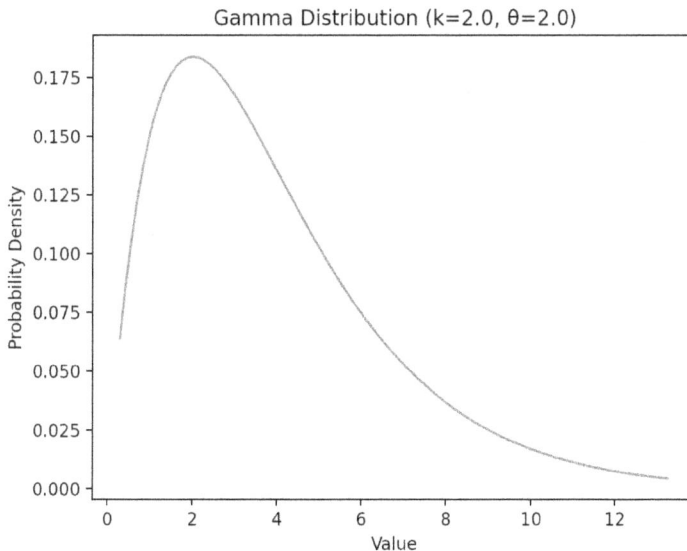

FIGURE 5.8: A gamma distribution

The Gaussian Distribution

The *Gaussian distribution* is named after Karl F. Gauss, and it is also called the *normal distribution* or the *bell curve*. The Gaussian distribution is symmetric: the shape of the curve on the left of the mean is identical to the shape of the curve on the right side of the mean. As an example, the distribution of IQ scores follows a curve that is similar to a Gaussian distribution.

The significant features of a Gaussian distribution include its symmetry and its standard deviation, which determines the "spread" of the data in the distribution. In addition, the mean, median, and mode in a Gaussian distribution tend to have similar (close) values.

A Gaussian distribution is also related to the Central Limit Theorem, which states that the sampling distribution of the sample means approaches a normal distribution as the sample size increases. Listing 5.8 displays the content of `gaussian_distribution.py` that renders a Gaussian curve superimposed over a histogram that approximates a Gaussian distribution.

Listing 5.8: gaussian_distribution.py

```
import numpy as np
import matplotlib.pyplot as plt


# Set the mean and standard deviation
mu, sigma = 0, 0.1


# Generate the data using numpy
s = np.random.normal(mu, sigma, 1000)


# Create the histogram of the data
count, bins, ignored = plt.hist(s, 30, density=True)


# Plot the distribution curve
plt.plot(bins, 1/(sigma * np.sqrt(2 * np.pi)) *
         np.exp( - (bins - mu)**2 / (2 * sigma**2) ),
         linewidth=2, color='r')
plt.show()
```

Listing 5.8 starts with several `import` statements followed by initializing the scalar variables `mu` and `sigma` (described in the previous code sample). The next portion of Listing 5.8 initializes histogram-related variables `count`, `bins`, and `ignored`, and then invokes the `plot()` method to display a histogram. Launch the code in Listing 5.8, and you will see the Gamma distribution shown in Figure 5.9.

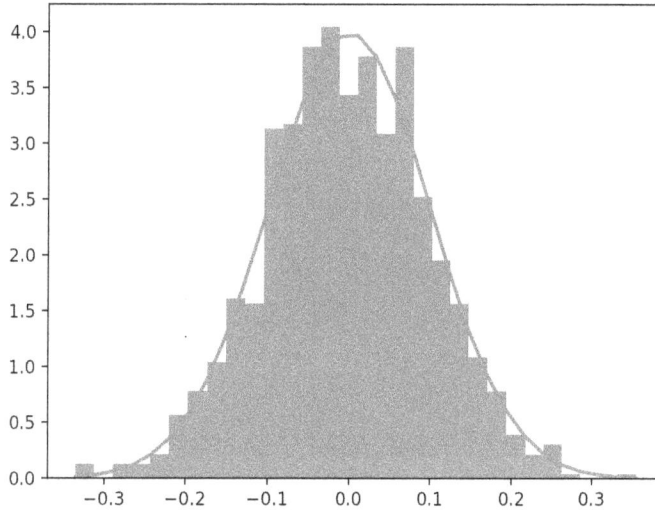

FIGURE 5.9: A Gaussian distribution

Listing 5.9 displays the contents of `gaussian_distribution2.py`, which is another code sample that renders the standard Gaussian (normal) distribution.

Listing 5.9: gaussian_distribution2.py

```
import numpy as np
import matplotlib.pyplot as plt
from scipy.stats import norm

# Normal distribution parameters
mu = 0    # mean
sigma = 1  # standard deviation
```

```
# Generate a range of x values
x = np.linspace(norm.ppf(0.01, mu, sigma),
                norm.ppf(0.99, mu, sigma),
                100)


# Generate the corresponding y values from the normal
distribution
pdf = norm.pdf(x, mu, sigma)


# Plot the distribution
plt.plot(x, pdf, label='norm pdf')
plt.title('Normal Distribution (µ=0, σ=1)')
plt.xlabel('Value')
plt.ylabel('Probability Density')


# Show the plot
plt.show()
```

Listing 5.9 starts with `import` statements and then initializes the scalar variables `mu` and `sigma` that represent the mean and standard deviation parameters for the normal distribution. The range of the `x` values is determined using the percent point function `ppf`, which is the inverse of the CDF. The `norm.pdf` function calculates the PDF of the normal distribution for those `x` values. Launch the code in Listing 5.9, and you will see the Gaussian distribution shown in Figure 5.10.

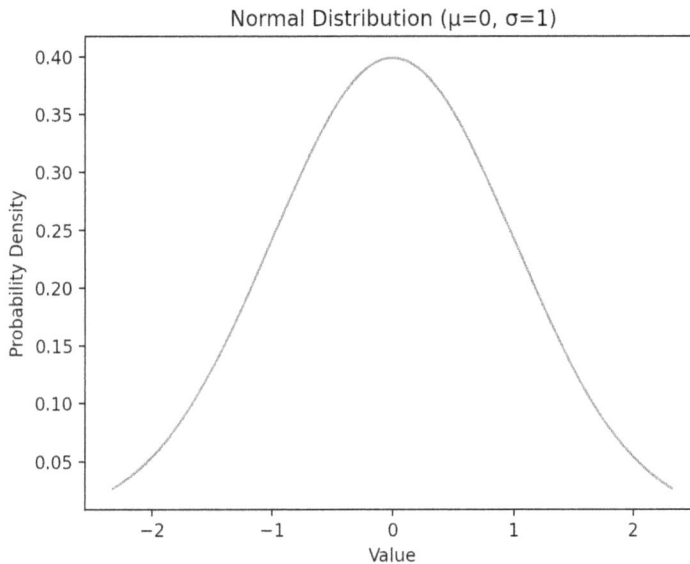

FIGURE 5.10: A Gaussian distribution

The Isotropic Gaussian Distribution

An *isotropic Gaussian distribution* is a Gaussian distribution whose covariance matrix equals the nxn identity matrix multiplied by sigma**2. Thus, the covariance is a diagonal matrix in which all entries are equal. Recall that the eigenvalues of a diagonal matrix are the entries in the diagonal: hence, an isotropic Gaussian distribution has n equal eigenvalues. (As a side note, when an eigenvalue appears multiple times in the set of eigenvalues for a covariance matrix, the number of occurrences of the eigenvalue is called the *multiplicity* of the eigenvalue.)

The Lognormal Distribution

A *lognormal distribution* is a continuous probability distribution of a random variable whose logarithm is normally distributed. Thus, if X is a random variable that is also lognormally distributed, then ln(X) is normally distributed.

You can use a log-transform on the target variable only when the distribution of the target variable is right-skewed. Generate a histogram plot to determine whether a target variable is right skewed by inspection. A right-skewed distribution occurs if the dataset contains outliers that cannot be excluded because they are significant for the model. There is an additional caveat: the target values must be non-negative to apply a log-transformation.

A log transformation of a right-skewed distribution tends to produce a more symmetric distribution. Contrary to what you might think, a log transformation of a left-skewed distribution tends to produce a distribution that is *more* left-skewed. Instead of a log transformation, use a power transformation for left-skewed distributions, such as squared values or performing some type of exponentiation.

Given a random variable X and a random variable Z with standard normal distribution, along with parameters mu and sigma, then the variable X has lognormal distribution if it has the following form:

```
x = exp(mu+sigma*Z)
```

More detailed information about the lognormal distribution, including various graphs, is accessible online:

https://en.wikipedia.org/wiki/Log-normal_distribution

Lognormal Graph

Listing 5.10 shows you a Python code sample that generates values from a lognormal distribution.

Listing 5.10: log_normal.py

```python
import numpy as np
import matplotlib.pyplot as plt
from scipy.stats import lognorm

# Lognormal distribution parameters
s = 0.954  # shape parameter
mu = 2.3   # location parameter

# Generate a range of x values
x = np.linspace(0.01, 10, 1000)

# Generate the corresponding y values from the log-
normal distribution
pdf = lognorm.pdf(x, s, scale=np.exp(mu))
```

```
# Plot the distribution
plt.plot(x, pdf, label='lognorm pdf')
plt.title('Log-normal Distribution (s=0.954, mu=2.3)')
plt.xlabel('Value')
plt.ylabel('Probability Density')

# Show the plot
plt.show()
```

Listing 5.10 starts with several `import` statements followed by initializing the variable `s` that represents the standard deviation of the log of the random variable, as well as the scalar variable `mu`, which is the mean of the log of the random variable.

The code snippet `np.linspace(0.01, 10, 1000)` is used to generate the x values over which we want to plot the lognormal distribution. Finally, the `lognorm.pdf` function calculates the PDF of the log-normal distribution for those x values. Launch the code in Listing 5.10, and you will see the lognormal distribution shown in Figure 5.11.

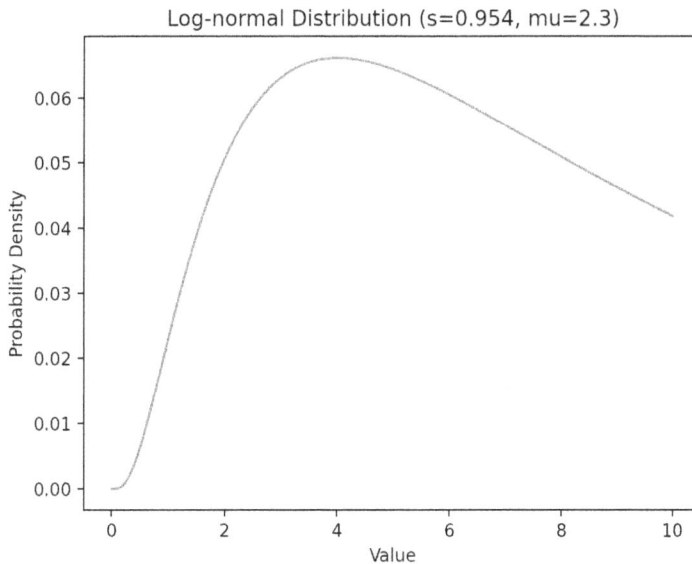

FIGURE 5.11: A lognormal distribution

Multiple Lognormal Graphs

Listing 5.11 shows you a Python code sample that uses a lognormal distribution from the SciPy open-source library. This code sample involves some familiarity with the NumPy open-source library, which is not discussed in this book (free online tutorials are available).

Listing 5.11: log_normal1.py

```
# pip3 install scipy
# pip3 install numpy
# pip3 install matplotlib

import numpy as np
import matplotlib.pyplot as plt
import scipy.stats as stats
from scipy.stats import norm

X = np.linspace(0, 8, 200)

# graph #1
stddev = 1
mean = 0
lognorm_dist = stats.lognorm([stddev], loc=mean)
lognorm_dist_pdf = lognorm_dist.pdf(X)

fig, ax = plt.subplots(figsize=(8, 5))
plt.plot(X, lognorm_dist_pdf, label="μ=0, σ=1")
ax.set_xticks(np.arange(min(X), max(X)))

# graph #2
stddev = 0.5
mean = 0
lognorm_dist = stats.lognorm([stddev], loc=mean)
```

```
lognorm_dist_pdf = lognorm_dist.pdf(X)
plt.plot(X, lognorm_dist_pdf, label="μ=0, σ=0.5")

# graph #3
stddev = 1.5
mean = 1
lognorm_dist = stats.lognorm([stddev], loc=mean)
lognorm_dist_pdf = lognorm_dist.pdf(X)
plt.plot(X, lognorm_dist_pdf, label="μ=1, σ=1.5")

plt.title("Log Normal Distribution")
plt.legend()
plt.show()
```

Listing 5.11 starts with four `import` statements followed by initializing the variable x with a range of values from 0 to 5,000 in multiples of 6. The majority of the code in Listing 5.11 consists of three similar blocks of code, each of which initializes the variables `stddev` and `mean`, and then plots the lognormal distribution of the variable X.

For example, the first such code block initializes the variables `stddev` and `mean` with the values of 0 and 1, respectively. Next, the variable `lognorm_dist` is initialized with the result of invoking the `lognorm()` method in the `stats` class. Then `lognorm_dist_pdf` is initialized by invoking the `pdf()` method of the variable `lognorm_dist`. After initializing the title and including the legend, the result is displayed with the `plt.show()` method. Launch the code in Listing 5.11, and you will see the image shown in Figure 5.12.

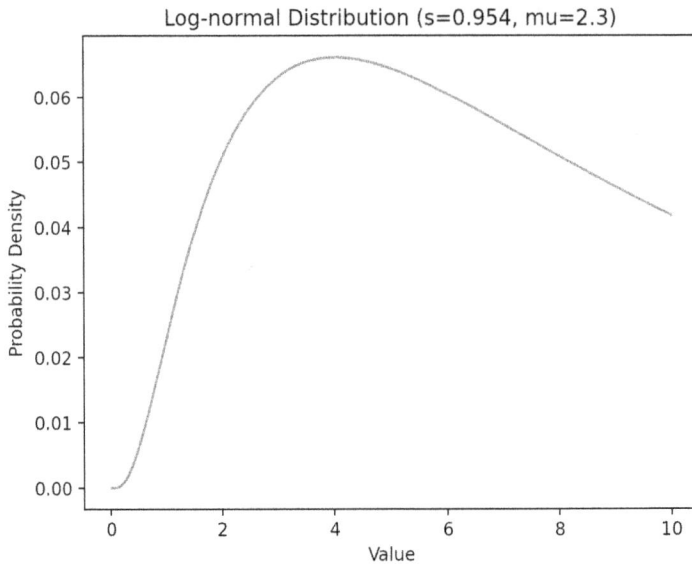

FIGURE 5.12: A lognormal distribution

The T-Distribution

A *t-distribution* is a probability distribution for which the population standard deviation is unknown and observations are taken from a normally distributed population. The t-distribution is used for estimating the significance of population parameters for small sample sizes or unknown variations. Although the t-distribution is also bell-shaped and symmetric, the t-distribution has "heavier" tails, which means that there is a greater chance for extreme values.

Listing 5.12: t_distribution.py

```
import numpy as np
import matplotlib.pyplot as plt
import scipy.stats norm, t

# Define the range of x values
x_values = np.linspace(-5, 5, 500)

# Define the mean and standard deviation
```

```
mean = 0
std_dev = 1

# Degrees of freedom for t-distribution
df = 3

# Calculate the PDF values for Gaussian distribution
gaussian_pdf = norm.pdf(x_values, mean, std_dev)

# Calculate the PDF values for t-distribution
t_pdf = t.pdf(x_values, df, mean, std_dev)

# Plotting
plt.figure(figsize=(12, 8))

# Plot Gaussian distribution
plt.plot(x_values, gaussian_pdf, label='Gaussian
(Normal) Distribution', linestyle='--')

# Plot t-distribution
plt.plot(x_values, t_pdf, label=f't-Distribution
(df={df})')



# Add title, labels, and legend
plt.title('t-Distribution vs Gaussian Distribution')
plt.xlabel('x')
plt.ylabel('Density')
plt.legend()
plt.grid(True)
plt.show()
```

Listing 5.12 starts with importing the required libraries: NumPy for generating x-values, `matplotlib.pyplot` for plotting, and `scipy.stats` for calculating the PDF and CDF of the Gaussian distribution as well as the t-distribution. Next, we initialize the scalar variables `mean`, `std_dev`, and `def` with the values 0, 1, and 3, respectively, for the mean, standard deviation, and the degrees of freedom `df` for the t-distribution (which you can modify to suit your needs). Then we invoke the `stats.t.pdf()` and `stats.t.cdf()` functions to calculate the PDF values for the Gaussian and t-distribution. The final code block in Listing 5.12 specifies appropriate labels, titles, and grids. Launch the code in Listing 5.12, and you will see the Gaussian distribution and the t-distribution shown in Figure 5.13.

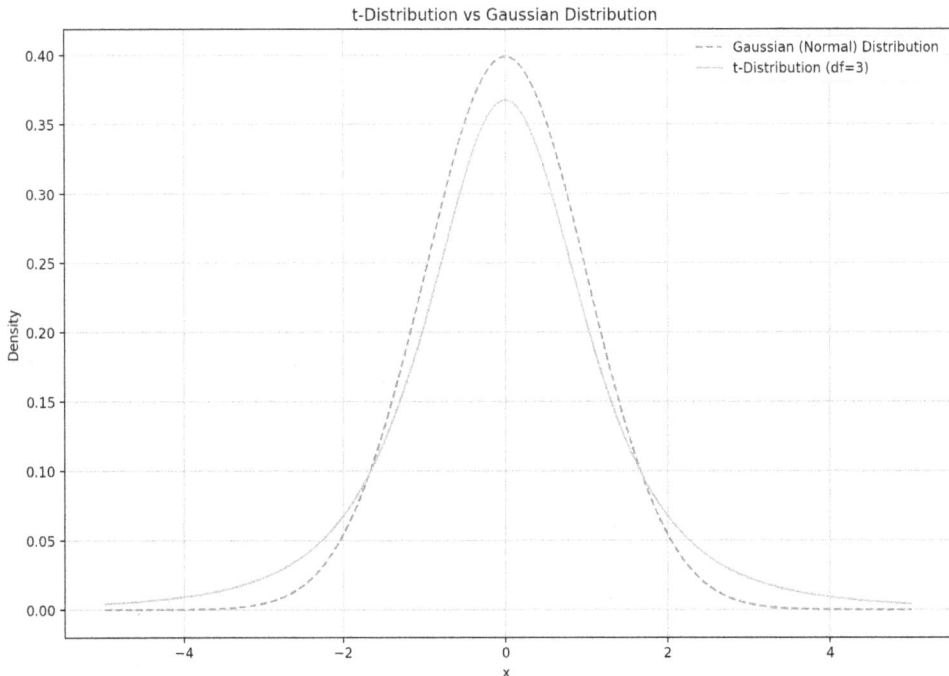

FIGURE 5.13: A t-distribution

The Uniform Distribution

A *uniform distribution* is a probability distribution for which all outcomes are equally likely. For example, tossing a fair coin or throwing a well-balanced die both have a uniform distribution. Another example involves a 52-card deck: the probability of drawing a card from each of the four suits is equally likely (1/4). In addition, a uniform distribution can be discrete as well as continuous.

If a uniform distribution is defined on the interval `[a,b]`, where `a < b`, then the mean and variance are as follows:

```
mean = (a+b)/2 and variance = (b-a)^2/12
```

Listing 5.13: uniform_distribution.py

```python
import numpy as np
import matplotlib.pyplot as plt
from scipy.stats import uniform

# Uniform distribution parameters
start = 0  # starting point of the distribution
width = 1  # width of the distribution

# Generate a range of x values
x = np.linspace(start, start + width, 100)

# Generate the corresponding y values from the uniform
distribution
pdf = uniform.pdf(x, start, width)

# Plot the distribution
plt.plot(x, pdf, label='uniform pdf')
plt.title('Uniform Distribution (start=0, width=1)')
plt.xlabel('Value')
plt.ylabel('Probability Density')

# Show the plot
plt.show()
```

In this example, start and width represent the parameters for the uniform distribution. The uniform distribution is a continuous distribution where all outcomes are equally likely within a given interval, specified by the start and width parameters. The uniform.pdf function calculates the PDF of the uniform distribution for those x values. Launch the code in Listing 5.13, and you will see the uniform distribution shown in Figure 5.14.

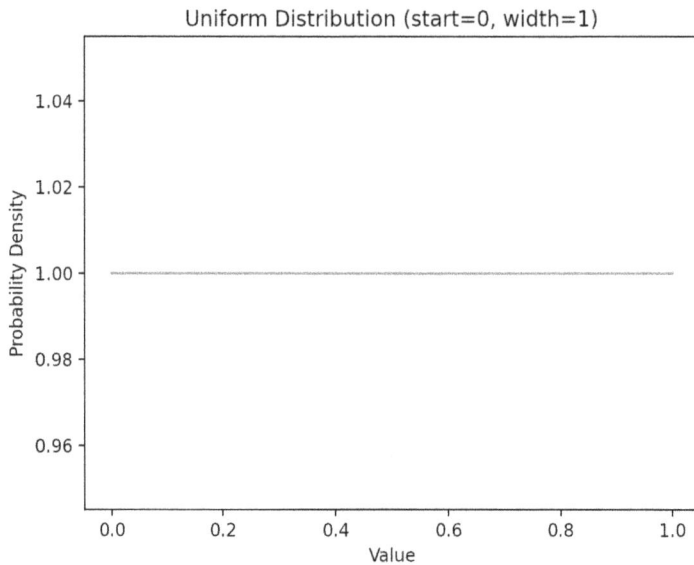

FIGURE 5.14: A uniform distribution

ADVANCED PROBABILITY FUNCTIONS

In addition to the probability functions that you have already seen in this chapter, there are some advanced continuous probability functions that require deeper mathematics, some of which are listed here:

* beta distribution
* zeta functions

The preceding functions are beyond the scope of this book, but you can perform an online search for articles that provide in-depth explanations for them. For your convenience, this section contains graphs of the preceding distributions.

Listing 5.14 displays the content of `beta_distribution.py` for rendering a beta distribution.

Listing 5.14: beta_distribution.py

```python
import numpy as np
import matplotlib.pyplot as plt
from scipy.stats import beta

# Define the range of x values
x_values = np.linspace(0, 1, 500)

# Define different shape parameters (alpha, beta)
shape_params = [(0.5, 0.5), (1, 1), (2, 2), (2, 5),
(5, 2)]

# Plotting
plt.figure(figsize=(12, 8))

for alpha, beta_ in shape_params:
    y_values = beta.pdf(x_values, alpha, beta_)
    plt.plot(x_values, y_values, label=f'Alpha={alpha},
Beta={beta_}')

plt.title('Beta Distributions with Different Shape
Parameters')
plt.xlabel('x')
plt.ylabel('Density')
plt.legend()
plt.grid(True)
plt.show()
```

Listing 5.14 starts with several `import` statements followed by initializing the variable `x_values` as an array of equally-spaced points between 0 and 1. The next code snippet initializes the variable `shape_params` with tuples of different shape parameters (α, β) for the beta distribution. The code will generate a beta distribution curve for each pair of shape parameters.

The next code snippet initializes a new figure for plotting with the dimensions of 12x8 inches, followed by a loop that iterates through the elements of `shape_params`. During each iteration, the calculations are made of the PDF of the beta distribution for the given α and β at each x value, followed by a code snippet that plots the curve for each set of shape parameters and labels it accordingly. Figure 5.15 shows the graph of several beta distributions.

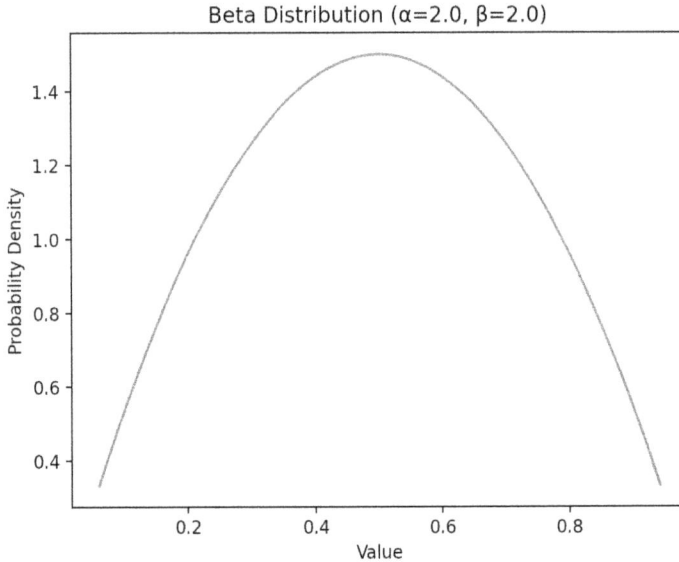

FIGURE 5.15: A beta distribution

Listing 5.15 displays the content of `zeta_distribution.py` for rendering a beta distribution.

Listing 5.15: zeta_distribution.py

```
import numpy as np
import matplotlib.pyplot as plt
#from scipy.stats import zetac
from scipy.special import zeta

# Define the range of x values
x_values = np.linspace(1.01, 10, 400)

# Calculate the corresponding zeta function values
zeta_values = zeta(x_values)
```

```
# Plotting
plt.figure(figsize=(10, 6))
plt.plot(x_values, zeta_values, label='Zeta Function')
plt.title('Zeta Function')
plt.xlabel('x')
plt.ylabel('Zeta(x)')
plt.grid(True)
plt.legend()
plt.show()
```

Listing 5.15 starts with several `import` statements followed by initializing the variable `x_values` that specifies a range of values from 1.01 to 10.0 that are used for generating zeta values with this code snippet:

```
zeta_values = zeta(x_values)
```

The remainder of Listing 5.15 specifies graph-related attributes, such as the title, horizontal label, and vertical label. Figure 5.16 shows the graph of a zeta distribution.

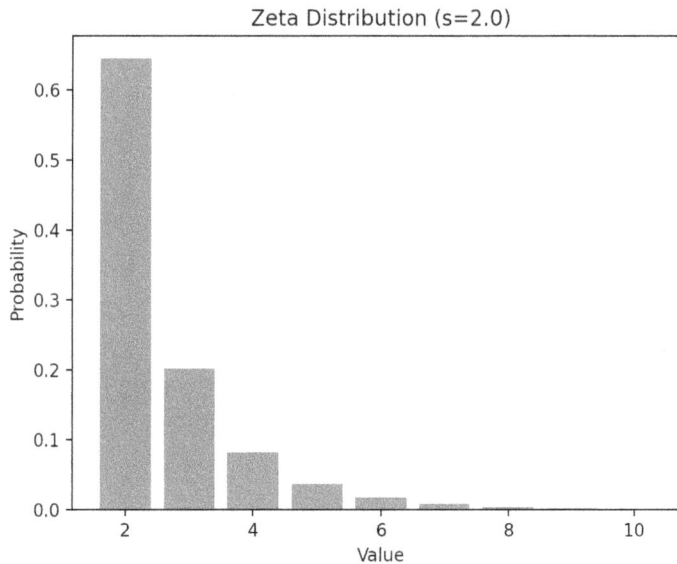

FIGURE 5.16: A zeta distribution

NON-GAUSSIAN DISTRIBUTIONS

Although you have already seen non-Gaussian distributions (such as the Poisson distribution), this section is included to describe some of the common features of non-Gaussian distributions.

Skewed Versus Uniform Distributions

The Gaussian distribution is a symmetric distribution, which means that the left half is a mirror image of the right half, where the line of symmetry is the vertical line that passes through the mean of the distribution. For example, a uniform distribution has identical probabilities for all possible outcomes, and therefore it is symmetric with respect to its mean.

By contrast, a *skewed* distribution is asymmetric, so one side is longer than the other side. Moreover, the skew of a distribution is known as the third moment, and kurtosis is the fourth moment.

An extensive list of non-Gaussian distributions (along with some explanations) can be found online:

https://en.wikipedia.org/wiki/List_of_probability_distributions

Data Types with Non-Normal Distributions

Many data types follow a non-normal distribution in nature, some of which are listed here:

- binomial distribution
- exponential distribution
- lognormal distribution
- largest-extreme-value distribution
- Poisson distribution
- Weibull distribution
- The data for the distributions in the preceding list are handled via techniques that are specific for non-normal data.

THE BEST-FITTING DISTRIBUTION FOR DATA

The Python-based open-source package distfit fits almost 90 different probability density distributions for univariate distributions against a dataset to determine the best-fitting distribution, and its home page is at the following URL:

https://pythonfoss.com/repo/erdogant-distfit-python-package-for-probability-density-function-fitting-of

Install distfit from the command line with this command:

```
pip3 install distfit
```

Listing 5.16 displays the content of the Python file distfit1.py that shows you how to use distfit for a set of randomly generated numbers to determine the best-fitting distribution.

Listing 5.16: distfit1.py

```
import matplotlib.pyplot as plt
import numpy as np
from distfit import distfit


# Generate 20000 normal distribution samples with mean
0, std dev of 4
X = np.random.normal(0, 4, 20000)


# Initialize distfit
dist = distfit()


# Determine best-fitting probability distribution for
data
dist.fit_transform(X)


dist.plot()
plt.show()
```

Listing 5.16 starts with import statements followed by the initialization of the variable X as a NumPy array of random numbers from a Gaussian distribution. Next, the variable dist is initialized with the result of invoking the distfit() method, after which its fit_transform() method is invoked with the array X. The result is then plotted and displayed. Launch the code in Listing 5.16, and you will see the following output:

```
[distfit] >fit..
[distfit] >transform..
[distfit] >[norm      ] [0.00 sec] [RSS: 0.0000857]
[loc=0.033 scale=3.988]
[distfit] >[expon     ] [0.00 sec] [RSS: 0.0796273]
[loc=-15.991 scale=16.024]
[distfit] >[pareto    ] [0.39 sec] [RSS: 0.0762562]
[loc=-938400416.619 scale=938400400.628]
[distfit] >[dweibull  ] [0.07 sec] [RSS: 0.0016850]
[loc=0.031 scale=3.421]
[distfit] >[t         ] [0.23 sec] [RSS: 0.0000857]
[loc=0.033 scale=3.988]
[distfit] >[genextreme] [0.55 sec] [RSS: 0.2325884]
[loc=15.967 scale=5.283]
[distfit] >[gamma     ] [0.20 sec] [RSS: 0.0001005]
[loc=-328.705 scale=0.048]
[distfit] >[lognorm   ] [0.40 sec] [RSS: 0.0001172]
[loc=-313.015 scale=313.012]
[distfit] >[beta      ] [0.21 sec] [RSS: 0.0000763]
[loc=-197.609 scale=319.085]
[distfit] >[uniform   ] [0.00 sec] [RSS: 0.0605828]
[loc=-15.991 scale=33.863]
[distfit] >[loggamma  ] [0.19 sec] [RSS: 0.0000748]
[loc=-648.814 scale=100.535]
[distfit] >Compute confidence interval [parametric]
```

Figure 5.17 shows the best-fitting distribution, determined by the dist-fit package, for a set of random numbers.

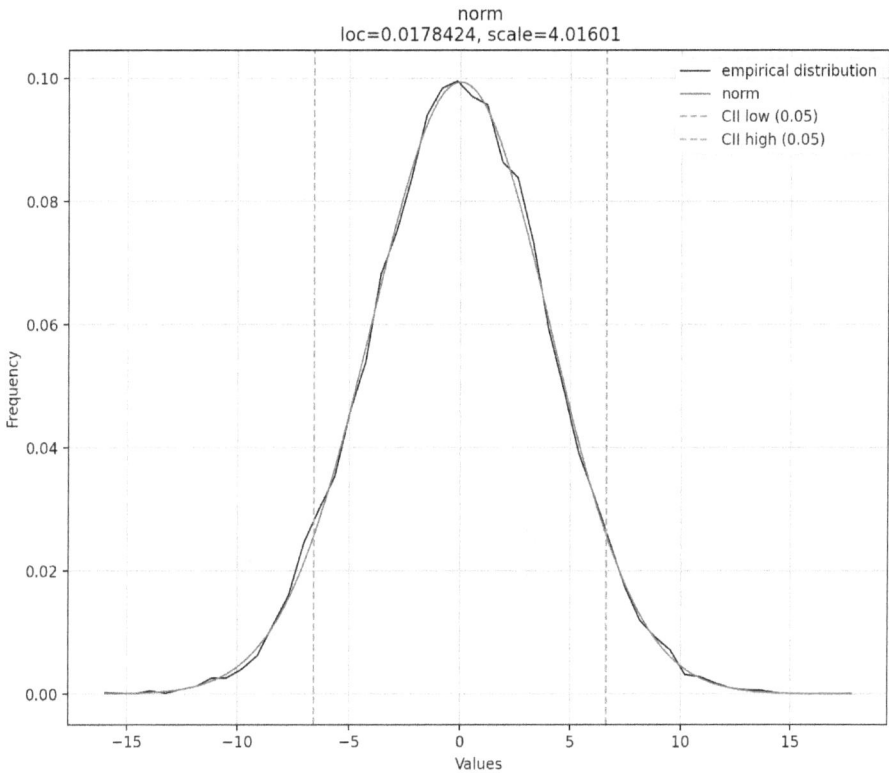

FIGURE 5.17: A best-fitting distribution for a set of random values

Listing 5.17 displays the content of the Python file `distfit2.py` that shows you how to use `distfit` for the `age` feature and then for the `fare` feature of the Titanic dataset to determine the best-fitting distribution.

Listing 5.17: distfit2.py

```
import matplotlib.pyplot as plt
import numpy as np
import pandas as pd
from distfit import distfit

df = pd.read_csv("titanic2.csv")
# part 1:
X = df['age']
```

```
# Initialize distfit
dist = distfit()

# Determine best-fitting probability distribution for
data
dist.fit_transform(X)

dist.plot()
plt.show()

# part 2:
X = df['fare']

# Initialize distfit
dist = distfit()

# Determine best-fitting probability distribution for
data
dist.fit_transform(X)

dist.plot()
plt.show()
```

Listing 5.17 starts with `import` statements followed by the initialization of the Pandas data frame `df` with the contents of the CSV file `titanic2.csv`. Next, the variable `X` is initialized with the contents of the `age` feature of `df`, and then `dist` is initialized with the result of invoking the `distfit()` method so that its `fit_transform()` method can be invoked. The result is then displayed.

The next portion of Listing 5.17 is similar: it initializes the variable `X` with the contents of the `fare` feature of `df`, and then `dist` is initialized with the result of invoking the `distfit()` method so that its `fit_transform()` method can be invoked. The result is then displayed. Launch the code in Listing 5.17, and you will see the following output:

```
[distfit] >fit..
[distfit] >transform..
```

```
[distfit] >[norm      ] [0.00 sec] [RSS: 0.0031262]
[loc=35.623 scale=15.629]
[distfit] >[expon     ] [0.00 sec] [RSS: 0.0085079]
[loc=0.920 scale=34.703]
[distfit] >[pareto    ] [0.02 sec] [RSS: 0.0232669]
[loc=-1.632 scale=2.552]
[distfit] >[dweibull  ] [0.01 sec] [RSS: 0.0033494]
[loc=34.308 scale=13.547]
[distfit] >[t         ] [0.04 sec] [RSS: 0.0031263]
[loc=35.623 scale=15.625]
[distfit] >[genextreme] [0.11 sec] [RSS: 0.0030927]
[loc=29.938 scale=15.527]
[distfit] >[gamma     ] [0.05 sec] [RSS: 0.0031233]
[loc=-1638.461 scale=0.146]
[distfit] >[lognorm   ] [0.03 sec] [RSS: 0.0261901]
[loc=0.920 scale=1.514]
[distfit] >[beta      ] [0.02 sec] [RSS: 0.0031099]
[loc=-27.402 scale=130.950]
[distfit] >[uniform   ] [0.00 sec] [RSS: 0.0063183]
[loc=0.920 scale=79.080]
[distfit] >[loggamma  ] [0.05 sec] [RSS: 0.0031307]
[loc=-3464.035 scale=503.956]
[distfit] >Compute confidence interval [parametric]
[distfit] >plot..
```

Figure 5.18 shows the best-fitting distribution, determined by the dist-fit package, for the age feature of the Titanic dataset.

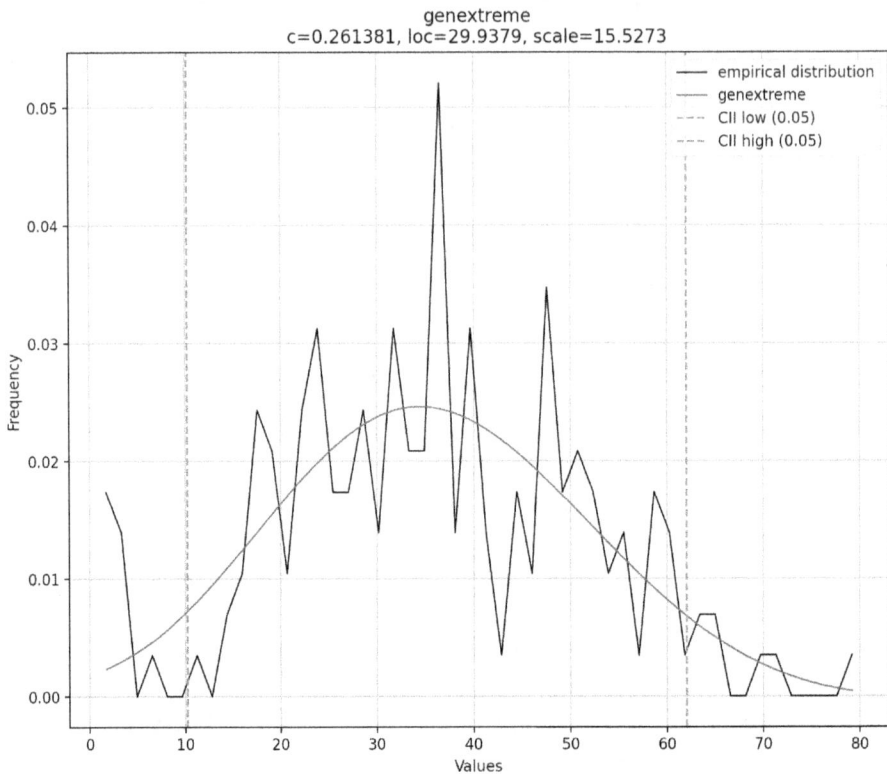

FIGURE 5.18: A best-fitting distribution for the "age" feature of the Titanic dataset

The second part of the output that is generated from Listing 5.17 is shown here:

```
[distfit] >fit..
[distfit] >transform..
[distfit] >[norm      ] [0.00 sec] [RSS: 0.0003880]
[loc=78.920 scale=76.280]
[distfit] >[expon     ] [0.00 sec] [RSS: 0.0002966]
[loc=0.000 scale=78.920]
[distfit] >[pareto    ] [0.03 sec] [RSS: 0.0011240]
[loc=-2.379 scale=2.379]
[distfit] >[dweibull  ] [0.03 sec] [RSS: 0.0003063]
[loc=53.100 scale=46.684]
[distfit] >[t         ] [0.06 sec] [RSS: 0.0002825]
[loc=57.571 scale=35.871]
```

```
[distfit] >[genextreme] [0.15 sec] [RSS: 0.0002238]
[loc=42.539 scale=35.032]

[distfit] >[gamma      ] [0.05 sec] [RSS: 0.0002313]
[loc=-0.627 scale=56.537]

[distfit] >[lognorm   ] [0.06 sec] [RSS: 0.0008034]
[loc=-0.000 scale=3.863]

[distfit] >[beta      ] [0.08 sec] [RSS: 0.0002318]
[loc=-0.612 scale=645627571382809.125]

[distfit] >[uniform   ] [0.00 sec] [RSS: 0.0006744]
[loc=0.000 scale=512.329]

[distfit] >[loggamma  ] [0.05 sec] [RSS: 0.0003968]
[loc=-27488.310 scale=3602.140]

[distfit] >Compute confidence interval [parametric]
[distfit] >plot..
```

Figure 5.19 displays the best-fitting distribution, determined by the distfit package, for the fare feature of the Titanic dataset.

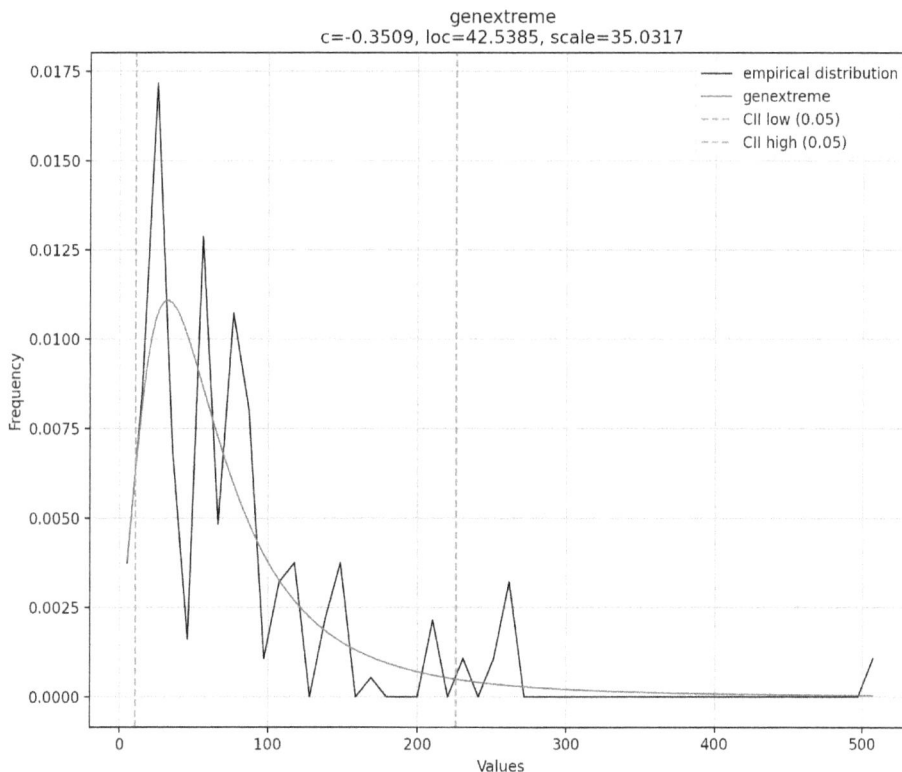

FIGURE 5.19: A best-fitting distribution for the "fare" feature of the Titanic dataset

SUMMARY

This chapter started with a brief description regarding the need for statistics, followed by some well-known terminology, and then the concept of a random variable, which can be either continuous or discrete.

Moreover, you learned about several discrete probability distributions, such as the Bernoulli, binomial, and Poisson distributions. Then you learned about continuous probability distributions, such as the chi-squared, Gaussian, and uniform distributions.

## CHAPTER 6

HYPOTHESIS TESTING

This chapter introduces you to an assortment of topics, including hypothesis testing and confidence intervals.

The first section introduces hypothesis testing, the null hypothesis, and the alternate hypothesis. The second section discusses the t-test and the z-test, followed by a discussion of p-values and their relevance to a null hypothesis. You will also see Python code samples to convert z-scores into their corresponding p-values.

The third section discusses point estimation, confidence level, and confidence intervals, and it includes a Python code sample.

WHAT IS HYPOTHESIS TESTING?

Hypothesis testing involves analyzing representative sample data (i.e., a subset of a population) to make inferences about the entire population. Regardless of the field of interest (e.g., science and machine learning), performing experiments invariably requires a null hypothesis.

Let's briefly discuss the null hypothesis in the following section, after which we will return to a discussion of hypothesis tests.

The Null Hypothesis and the Alternate Hypothesis

In simple terms, the *null hypothesis* (H0) is any statement whose validity you want to test, whereas the *alternate hypothesis* (H1) is the negation of the null hypothesis. The null hypothesis H0 asserts that a given assumption (specified by you) is true, whereas the alternate assumption H1 asserts that the null hypothesis is false. Examples of the null hypothesis are listed here:

- The flow of traffic follows a Poisson distribution.
- The earth revolves around the sun.
- A single six-sided die is fair.

In addition, a null hypothesis can contain the arithmetic symbols =, !=, <, >, >=, or <= that represent equal, not equal, less than, greater than, greater than or equal, less than or equal, respectively. Simplified examples of null hypotheses with these symbols are shown here:

The mean weight of a small bag of potato chips = 7 ounces.
The mean weight of a small bag of potato chips != 7 ounces.
The mean weight of a small bag of potato chips < 7 ounces.
The mean weight of a small bag of potato chips > 7 ounces.
The mean weight of a small bag of potato chips == 7 ounces.
The mean weight of a small bag of potato chips <= 7 ounces.

You must specify a null hypothesis before you conduct an experiment, after which you perform tests in order to support or negate the null hypothesis. Make sure that you do not reverse the order: it is incorrect to conduct an experiment and then formulate a null hypothesis.

Although the goal of hypothesis testing is to either accept or reject the null hypothesis, rejecting the null hypothesis does *not* automatically accept the alternate hypothesis.

At this point we can "condense" hypothesis testing, as shown in the following list:

- formulating a null hypothesis
- formulating an alternate hypothesis
- type I errors (rejects the null hypothesis when it is true)
- type II errors (accepts the null hypothesis when it is false)

For more information, please read the following online articles:

https://en.wikipedia.org/wiki/Statistical_hypothesis_testing
https://lih-verma.medium.com/hypothesis-testing-p-value-z-test-and-t-test-fd1a643120c0

Statistically Significant Results

We need a mechanism by which we can accept or reject the null hypothesis. Doing so involves evaluating the outcome of a given test on the sample data that we have for our experiment. An outcome or a result is considered a *statistically significant result* if it is an outcome that enables you to reject the null hypothesis. A statistically significant result has a very low probability of occurrence if the null hypothesis is true; consequently, that result is "associated" with a rejection of the null hypothesis.

With the preceding points in mind, let's discuss the major types of hypothesis tests that are available.

Two Types of Hypothesis Tests

There are two major types of hypothesis tests: *parametric tests* and *non-parametric* tests. Parametric tests are suitable for samples from a normally distributed population, whereas non-parametric tests are suitable for samples from distributions in which the distribution is unknown. Parametric tests can be done in non-normal settings if the underlying distribution is known.

The following link provides a flow chart to assist you in making a decision regarding the test(s) to perform:

https://www.kdnuggets.com/2021/09/hypothesis-testing-explained.html

A tutorial on how to use `easy-ht` is available online:

https://pypi.org/project/easy-ht/

Now that you have an understanding of the null hypothesis and the two major types of hypothesis testing, let's look at the major sub-components of hypothesis tests.

COMPONENTS OF HYPOTHESIS TESTING

Let's expand on the short list of steps for hypothesis testing that you learned in a previous section. Specifically, the major building blocks of hypothesis testing are shown here:

- hypotheses H0 and H1
- statistical tests
- probability distributions
- test statistics
- critical values
- level of significance (alpha)
- p-value

Note that rejecting the null hypothesis requires gathering sufficient evidence that enables you to do so. The alternate hypothesis refutes (rejects) the null hypothesis and asserts that some *other* hypothesis is correct. However, there can be *multiple* alternate hypotheses, and as noted earlier, rejecting the null hypothesis does not automatically guarantee that your alternate hypothesis is true.

For instance, one null hypothesis can be the statement that the mean weight of a small bag of potato chips = 7 ounces. If this hypothesis is false, then one (but not both) of the following alternate hypotheses is true:

The mean weight of a small bag of potato chips > 7 ounces.
The mean weight of a small bag of potato chips < 7 ounces.

The second item in our list, statistical tests, pertains to probability distributions, some of which are discussed in the previous chapter. (Critical values, level of significance (alpha), and p-values are discussed later in this chapter.)

What to Remember about Hypothesis Testing

The purpose of hypothesis tests is to test properties of population parameters indirectly through testing properties of the sample statistics. Population parameters are unknown, but we *do* know the statistical properties of samples, which is the rationale for obtaining multiple samples from a population.

Moreover, since each sample from a population is different, the point estimate for each sample has a margin of error. A sample distribution is created by repeatedly selecting samples from a population. As you know from the CLT, the set of mean values of samples can be approximated by a Gaussian distribution. See the following article for more information:

https://navistats.medium.com/the-usual-misinterpretations-in-statistical-inference-2030e8c81e76

TEST STATISTICS

There are several popular and well-known tests that you can perform when you want to test the value of a statistic, two of which are listed here:

- z-test
- t-test

The following subsections provide additional details regarding the tests in the preceding list.

What is a z-test and a z-score?

The area under the Gaussian distribution of the region that is bounded by one standard deviation to the left of the mean and one standard deviation to the right of the mean equals approximately 68% of the total area under this curve.

A region that is two standard deviations to the left of the mean and to the right of the mean has an area that is approximately 95% of the total area. The region that is bounded by 3 standard deviations has an area that is approximately 99.7% of the total area under the standard normal distribution.

Given a percentage P, the z-score is the distance from the left of the mean as well as the right of the mean such that the area of that region equals P. From the preceding paragraph, you know the following values:

- A z-score of 1 corresponds to the value 68% for P.
- A z-score of 2 corresponds to the value 95% for P.
- A z-score of 3 corresponds to the value 99.7% for P.

Moreover, given a number X, its z-score is computed as follows, where the mean and `sigma` are the mean and standard deviation of a Gaussian distribution:

```
z-score = (X-mean)/sigma
```

Using a z-score or a t-score

This section contains a bullet list of items that describe when to use a z-score and when to use a t-score.

Use a z-score when:

- You know the population variance or standard deviation.
- The sample size is large (usually n>30n>30, but this is a rule of thumb).
- The data is approximately normally distributed or the sample size is large enough to invoke the central limit theorem for sample means.

Use a t-score when:

- You do NOT know the population variance or standard deviation, so you have to estimate it from your sample.
- The sample size is small (typically n≤30n≤30).
- The data is approximately normally distributed or you're working with sample means and the sample size is not large enough to invoke the central limit theorem.
- You're performing a Student's t-test.

WORKING WITH P-VALUES

A *p-value* is a numeric value that is used to determine whether the null hypothesis will be rejected. Remember that a small p-value (usually less than 0.05) enables you to reject the null hypothesis, whereas a larger p-value (greater than 0.05) allows you to refuse to reject the null hypothesis, and a p-value of 0.05 is inconclusive with respect to the null hypothesis.

The p-value is the probability of obtaining a sample with the current properties or more extreme. Consequently, a very low p-value indicates that the null hypothesis is likely to be incorrect. The p-value associated with hypothesis tests enables us to make decisions about the null hypothesis when the p-value is small (e.g., 0.05, 0.01, or 0.005)

Recall that the standard deviation is a measure of how far the data values are from the mean, whereas the standard error measures the accuracy of the mean of a sample in comparison to the true population mean.

As a reminder, the p-value is used to reject the null hypothesis if the p-value is less than the significance level, which is the "cut-off" value. The p-value varies from sample to sample, and typical values for the significance level include 0.01 and 0.05.

In addition, we use t-statistics for accepting or rejecting H0. There are three common ranges of values that helps you decide the significance of results after a hypothesis test:

- p-value > 0.05 (cannot reject the null hypothesis)
- p-value = 0.05 (impossible to accept or reject the null hypothesis)
- p-value < 0.05 (reject the null hypothesis)

The p-value and significance level are prone to misinterpretation. A significance level of 0.05 means the probability of a type I error is 5%. That is, if the null hypothesis is true, then 5% of the time it will be rejected due to randomness. Thus, a p-value of 0.05 does not mean that the probability of the null hypothesis is 5%.

Alpha Value and P-Value

An *alpha value* is also called the *significance level*, which was discussed in the previous section. If the probability of getting the sample score, or a more extreme value, is less than alpha, then it is deemed significantly different from the population (or possibly belonging to some new sample distribution).

For example, suppose the p-value is less than alpha, where alpha equals 0.05 (for this example). This indicates that there is a probability of less than 5% that the result occurred purely by chance.

Consider the following scenario. Suppose you conduct a study to test the effectiveness of a drug. Before starting the study, you set $\alpha=0.05$, and after conducting the study and performing statistical analysis, you determine a p-value of 0.03.

Since $p=0.03 < \alpha$ (=0.05), you reject the null hypothesis and conclude that the drug is effective. Thus, the p-value and α work together to help you determine the statistical significance of your test results.

What is the Correct Alpha Value?

Many online articles specify either 0.01 or 0.05 as the "cut-off" value for accepting or rejecting the null hypothesis. According to the following online paper, the value 0.05 still allows for false positives, and the authors recommend the value 0.005 instead of 0.05 for the alpha value. More details are available online:

https://imai.fas.harvard.edu/research/files/significance.pdf

Consequently, it is worthwhile to experiment with different alpha values, such as 0.005, 0.01, and 0.05, and then compare the results.

One more point to note: the alpha value measures the probability of type I errors, whereas beta measures type II errors, which equals the probability `P(insignificance|H1)`. By contrast, a type II error occurs when we fail to reject the null hypothesis when the null is false. The probability equals `P(Accept H0 | H0 is False)`. Recall that type I errors are false positives, whereas type II errors are false negatives.

Calculating a P-Value from a Z-Score

Listing 6.1 displays the content of `zscore2pvalue.py` that shows you a set of z-scores (discussed earlier) and their corresponding p-values. The p-values in this example are associated only with "<=" types of tests, whereas p-values for ">=" types of tests would equal 1 minus the computed results.

Listing 6.1: zscore2pvalue.py

```
import scipy.stats as stats

scores = [-5,-4,-3,-2,-1,0,1,2,3,4,5,10,100]
```

```
for score in scores:
  pvalue = stats.norm.cdf(score)
  print(f"z-score: {score:3d} p-value: {pvalue}")
```

Listing 6.1 starts with an `import` statement and then initializes the variable `scores` with a list of integer values. The next portion of Listing 6.1 is a loop that iterates through the values in the `scores` variable, computes its associated `p-value`, and then prints both the score and the associated `p-value`. Launch the code in Listing 6.1, and you will see the following output:

```
z-score:  -5 p-value: 2.8665157187919333e-07
z-score:  -4 p-value: 3.167124183311986e-05
z-score:  -3 p-value: 0.0013498980316300933
z-score:  -2 p-value: 0.022750131948179195
z-score:  -1 p-value: 0.15865525393145707
z-score:   0 p-value: 0.5
z-score:   1 p-value: 0.8413447460685429
z-score:   2 p-value: 0.9772498680518208
z-score:   3 p-value: 0.9986501019683699
z-score:   4 p-value: 0.9999683287581669
z-score:   5 p-value: 0.9999997133484281
z-score:  10 p-value: 1.0
z-score: 100 p-value: 1.0
```

As you can see in the preceding output, the `z-scores` are symmetric with respect to the vertical axis, whereas the `p-values` are *not* symmetric. For example, the `p-value` for a z-score value of n is different from the `p-value` for a z-score value of -n.

Notice that if m and n are two z-scores where $m < n$ and $n < 10$, then the `p-value` for m is strictly less than the `p-value` for n. For z-scores that are 10 or larger, the corresponding `p-value` equals 1.

In fact, the `p-values` are *monotonically increasing* for `z-scores` less than 10, because they are cumulative values whose maximum value is 1, similar to a probability density function for a continuous random variable.

WORKING WITH ALPHA VALUES

An *alpha value* is another numeric value that is used for accepting or not rejecting a null hypothesis. As you have seen in previous subsections, a p-value and an alpha value are expressed as decimal values. Both values are between 0 and 1: a p-value is a computed value whereas an alpha value tends to be 0.05 or less.

Alpha Value in Conjunction with a P-Value

An *alpha value* is a threshold value that is used in conjunction with a p-value. For example, if alpha = 5%, then we accept a 5% chance that we are wrong when we reject the null hypothesis. Think of the alpha value as a "cut-off" point for deciding whether to reject (or not reject) the null hypothesis. In other words, hypothesis terminology is based on rejection versus failure-to-reject instead of acceptance versus non-acceptance.

For example, suppose the alpha value is 5% and the probability of occurrence of a particular outcome of a test is *less than* 5%: in this case, *reject* the null hypothesis. Conversely, if the probability of occurrence of a particular outcome of a test is *greater than* 5%, then *do not reject* the null hypothesis.

Based on the preceding statement, *outcomes whose probability is greater than the alpha value give us greater confidence in the validity of the null hypothesis.* However, *outcomes whose probability is less than the alpha value suggest that it is more likely that such outcomes are not due to chance.* Hence, the null hypothesis is less likely to be true.

POINT ESTIMATION, CONFIDENCE LEVEL, AND CONFIDENCE INTERVALS

Point estimation estimates a population parameter with a single number using techniques such as the method of moments and maximum likelihood estimator.

Next, an *interval estimate* provides a range of plausible values for an unknown parameter instead of a single point estimate. This range is calculated from a given set of sample data and is intended to offer a degree of confidence that the parameter falls within that range. The confidence level, usually expressed as a percentage like 95% or 99%, indicates how confident we are that the interval contains the true parameter value.

A *confidence level* is the *percentage of times* you expect to reproduce an interval estimate between the upper and lower bounds that determine what is called a confidence interval.

A *confidence interval* is the range of values (i.e., an interval of the form [a,b]) that estimated values are expected to occur for a particular percentage of the time.

As an example, suppose you want to estimate the average weight of apples from a large orchard. You randomly select 50 apples and weigh them. The sample mean weight is 150 grams, and the sample standard deviation is 30 grams. Let's calculate the confidence interval CI in a three-step process, as shown here.

Step 1: Choose a Confidence Level

Suppose that you want a 95% confidence level. The z-value for a 95% confidence level in a standard normal distribution is approximately 1.96.

Step 2: Calculate the margin of error.

The margin of error (ME) can be calculated as follows:

```
ME = Z × (Standard Deviation)/(Sample Size)
```

Calculate the standard deviation as follows:

```
std = 1.96×30/sqrt(50) ≈ 8.3 grams
```

Step 3: Calculate the confidence interval.

The confidence interval (CI) can be calculated as follows:

```
CI = Sample Mean±std
CI = 150±8.3
CI = [141.7,158.3]
```

With 95% confidence, you can say that the average weight of apples in the orchard falls between 141.7 grams and 158.3 grams.

A confidence interval is not the probability that a population parameter lies in a given interval. Instead, a confidence interval [a,b] involves a confidence level CI (specified by you) that has this property: there is a CI% probability that a population parameter (such as the mean) lies in the confidence interval [a,b].

Deriving the Confidence Interval Formula

Let's derive a confidence interval via a sequence of three intuitive steps. First, an `N(0,1)` distribution is a Gaussian distribution with a mean `mu` equal to 0 and standard deviation `std` equal to 1. In addition, the following intervals capture the approximate number of values in a Gaussian distribution [1]:

- `[mu-1*std, mu+1*std]`: approximately 68.27% of the values
- `[mu-2*std, mu+3*std]`: approximately 95.45% of the values
- `[mu-3*std, mu+3*std]`: approximately 99.73% of the values

For additional information, read the following online article:

https://en.wikipedia.org/wiki/Normal_distribution

Second, if we take into account the number of observations in the sample and call this value `N`, then a refinement of the first interval shown in the preceding list is here:

`[mu-std/`**`sqrt(N)`**`, mu+std/`**`sqrt(N)`**`]`

As N increases in value, the value of sqrt(N) also increases (more slowly, of course), and therefore the width of the preceding interval decreases because sqrt(N) appears as the denominator of the std value.

The third step modifies the interval in the preceding step by including a parameter `alpha` (which is the significance level) that results in a confidence interval that depends on `alpha` and `N`, as shown in the following result:

$$\left[\bar{x} - z_{\alpha \backslash 2} * \frac{\sigma}{\sqrt{n}} , \bar{x} + z_{\alpha \backslash 2} * \frac{\sigma}{\sqrt{n}} \right]$$

Since sigma is assumed to be 1, we could restate this as follows:

$$\left[\bar{x} - \frac{z_{\alpha \backslash 2}}{\sqrt{n}} , \bar{x} + \frac{z_{\alpha \backslash 2}}{\sqrt{n}} \right]$$

A Confidence Interval for a Standard Normal Distribution

This section contains an example of constructing a confidence interval when the value of sigma is known, which involves the following steps:

- Calculate the sample mean `mu` from the sample data.
- Determine the corresponding z-score.
- Calculate the value of EBM (error bound).
- Construct the confidence interval.

Recall the confidence interval whose formula was derived in the previous section, which is reproduced here for your convenience:

$$\left[\bar{x} - \frac{z_{\alpha \setminus 2}}{\sqrt{n}} \, , \, \bar{x} + \frac{z_{\alpha \setminus 2}}{\sqrt{n}} \right]$$

For simplicity, let's introduce the term *error bound* (EB): this is just the term in the preceding interval that we subtract as well as add to the value of `xbar`. Hence, we can rewrite the preceding interval as follows:

```
[xbar-EB, xbar+EB]
```

In addition, we will make the following assumptions regarding the population distribution:

- The population has a standard normal distribution.
- The confidence level (CL) is 95%.
- The value of `sigma` is 3.
- The value of `xbar` is 20.
- The value of `n` is 36.

The value of alpha is 5% (= 1 - 95%), and so we need to find the z-score `z1` for the value 0.025 (= `alpha/2`), which in turn equals 1.645. Now we can calculate the value of EB as follows:

```
EB = z1*sigma/sqrt(n) = 1.645 * 3/6 = 0.8225
```

We can construct the associated confidence interval by subtracting and adding the term `EB` from the mean, as shown here:

```
CI = [xbar-EB, xbar+EB]
   = [20-0.8225, 20+0.8225]
   = [19.1775, 20.8225]
```

The following links provide online calculators for finding confidence intervals as well as z-scores:

https://www.calculator.net/confidence-interval-calculator.html

https://www.calculator.net/z-score-calculator.html

Confidence Level Versus Significance Level

A statistical test that measures *significance* involves binary outcomes, such as yes or no, and reject or fail to reject. The *significance level* is defined as the probability of rejecting the null hypothesis when the null hypothesis is true and expressed via the following formula.

```
Significance level = 1 - Confidence level
```

Recall that a confidence interval is an interval of the form `[a,b]`, where a < b, such that the true value is somewhere in that interval with a *confidence level*. For example, a 95% confidence level indicates 95% certainty, and a 5% significance level indicates the risk of concluding that a difference exists when there is no actual difference.

WHAT IS A/B TESTING?

A very common use case for A/B testing involves Web pages. Specifically, an *A/B test* (also called a *split test*) for a Web page tries to determine which changes (if any) in the contents of that Web page that result in a different outcome.

The specific changes to a given Web page can be as simple as changing the location of an input field or rearranging the order in which large components are displayed. Examples of desired outcomes include more users registering for an account or longer periods of time spent on the Web page. An A/B test involves these steps:

1. Create a variation of the original Web page.

2. Each Web page receives 50% of Web traffic.

3. Analyze the differences in the results.

For more information, please visit the following URL:

https://towardsdatascience.com/typical-9-step-a-b-test-workflow-for-data-scientists-in-2022-d672c9a0d658

The alternate Web page differs from the original Web page (control) in any fashion that you want to test to determine whether the alternate Web page results in a statistically significant outcome. For example, you might be interested in the number of new users who register for an account through the alternate Web page. Another outcome could be the amount of time that users spend on the alternate Web page. Keep in mind that the results of an

A/B test might also depend on the device (e.g., phone versus laptop) because the distribution of users is skewed to one type of device.

In general, A/B testing involves hypothesis testing for a randomized experiment with two variables A and B. The purpose of A/B testing is to determine which of two options is better in terms of a given metric.

As an illustration, suppose a test of a Web page consists of variants A and B that differ by a single feature, and each variant receives 50% of traffic. Users are divided into two groups that are called the *control* group and the *test* group, and denoted group A and group B, respectively. Group A users are shown the old Web page and Group B users are shown the new Web page. Perform the test repeatedly until meaningful results are obtained: if an alternate page is statistically better performing (e.g., user conversion), then replace the original page with that page.

A/B testing works best for small and incremental changes rather than large or multiple changes. However, keep in mind that A/B testing can be time consuming, and tends to involve a large group for testing in order to gather meaningful results. See the following articles for more information:

https://euelfantaye.medium.com/a-b-testing-with-machine-learning-277da2750123
https://github.com/heavye/abtest-mlops

Sequential A/B Testing

The actions of users on these Web pages provides feedback that in turn is modeled via probability distributions called fA and fB that have the means mA and mB, respectively. The Web page with a percentage of users creating an account is deemed the more preferable version of the Web page. Users who see one Web page are usually unaware of the other Web page (and they do not need to know). The purpose is to ascertain which Web page performs "better" based on a specific criterion, such as revenue generation, the number of users who registered via a sign-up form, and so forth. (Most MarTech software provides a built-in A/B testing function.)

A high-level description of the main steps in an A/B testing process consists of the following set of steps:

- Formulate a hypothesis.
- Start the test and gather statistical information.
- Use the information to accept or reject the hypothesis.
- A stopping rule determines when the A/B test has been completed.
- A decision rule accepts/rejects H0 based on the available evidence.

Use the final set of data values to decide whether your hypothesis was 1) correct, 2) incorrect, or 3) inconclusive.

A/B Tests and Confidence Levels

In the context of A/B testing, the *confidence level* refers to the probability that the range of values (confidence interval) you have obtained contains the true value of the parameter you are estimating. In simpler terms, a confidence level gives you a measure of the reliability of the results of your A/B test.

For instance, if you have a confidence level of 95%, it means that if you were to repeat your experiment 100 times, you would expect the confidence interval to contain the true parameter value 95 times out of 100.

A high confidence level indicates that you can trust the results of your A/B test to a great extent. However, it is essential to note that a higher confidence level would generally require a larger sample size, and it also results in a wider confidence interval, implying less precision in your estimate.

Thus, choosing a confidence level involves balancing precision and reliability, and a confidence level of 95% is a standard choice in many statistical analyses, offering a good balance between these two factors.

Two important factors that affect a confidence level are the test sample size and the variability of the results. Some A/B testing software packages include the following columns:

* Confidence
* Statistical significance
* Significance

Typically, these columns contain percentages between 0 and 100%, which determines the statistical significance of the results.

A/B Testing Errors

Hypothesis testing (A/B testing) is a decision-making method in which there are several possible outcomes of the test:

* no error
* type I error
* type II error
* type S (sign) error
* type M (magnitude) error

No errors are obviously preferred, but it is possible to have either a type I error or a type II error. Read the following article for more information regarding Type I and Type II errors:

https://towardsdatascience.com/the-joy-of-a-b-testing-theory-practice-and-pitfalls-de58acbdb04a

Avoiding Statistical Errors

One technique for avoiding statistical errors is the use of random sampling. When we conduct an A/B test, visitors are distributed randomly among different variations. We use the results for each variation to judge how that variation will behave, if it is the only design visitors see.

Interleaving Algorithm

Given two ranked lists of items A and B, the interleaving algorithm selects items from A and B in an alternating fashion. For example, suppose that the items in A and B are ranked from highest to lowest:

```
S = {S1, S2, . . ., Sk}
T = {T1, S2, . . ., Tk}
```

Then the interleaving algorithm generates either the list C or D, depending on whether the first item is select from set S or from set T, respectively:

```
U = {S1, T1, S2, T2, . . ., Sk, Tk}
V = {T1, S1, T2, S2, . . ., Tk, Sk}
```

The interleaving algorithm has produced very good results for Netflix, where the lists A and B are lists of movie recommendations:

https://netflixtechblog.com/interleaving-in-online-experiments-at-netflix-a04ee392ec55
https://towardsdatascience.com/the-joy-of-a-b-testing-part-ii-advanced-topics-6c7f6cf71e4c

A/B/n Testing

A/B testing involves two versions of content (such as a Web page) in which traffic is divided equally, whereas A/B/n testing extends A/B testing by supporting multiple versions. For example, if you have three versions of a Web page, A/B/n tests provide one-third of the traffic to each of those three Web pages. Read the following article for more information:

https://medium.com/readwrite/a-b-n-and-mvt-testing-business-benefits-similarities-and-differences-feaa589c1106

Multivariate Testing (MVT)

MVT identifies the effectiveness of combinations of different elements on a website to determine which combination yields more conversions. Although MVT tests and A/B/n tests involve the same elements, MVT testing can produce more significant results than A/B/n testing.

Quasi-Experiments

In simple terms, a *quasi-experiment* is an experiment in which the creation of the control group and the treatment group is not completely random. The key idea is that a process that is "close enough" to random will achieve reasonable results.

However, quasi-experiments need to be finely constructed because they can produce unexpected results that infer causality even when it is not true.

THE LIFESPAN OF AN A/B TEST

The relevance of the results of an A/B test are difficult to ascertain because the results of an A/B test can (and often will) change over a period of time. Since you do not know when the results diminish, perform the A/B test several times at regular intervals, and then collect and compare the results. Repeat this process again after a longer period of time. The following factors that influence the frequency of performing the A/B test:

- the cost of performing the test
- the duration of the A/B test
- the time of year (seasonality)
- the demographics of the users

Interestingly, some users do not automatically accept new versions of a Web page, even if it is a better Web page, because they have become accustomed to the original. For example, email providers sometimes make significant changes to the user interface. However, users might resist upgrading to the latest interface for various reasons (e.g., they feel that there is too much re-learning associated with the new interface). Ironically, after they do perform the upgrade (perhaps the older interface has an imminent end-of-life), sometimes users actually *prefer* the newer interface after they have become comfortable with its design and features.

MAXIMUM LIKELIHOOD ESTIMATION (MLE)

The task of fitting a model to the data in a dataset involves some type of estimation, where the latter can be either point estimation or interval estimation. The Maximum Likelihood Estimation (MLE) is an example of point estimation, which involves finding parameters that provide the greatest likelihood that a set of values will actually occur.

The MLE enables us to estimate the values of the parameters of a probability distribution, which involves maximizing the value of a likelihood function that in turn equals the product of a set of probability densities. The MLE does not maximize the *size* of the estimator. Instead, the MLE maximizes the *probability* of the data. With the preceding observations in mind, suppose we have the following scenario:

- X1, X2, ..., Xn is a set of independent random variables.
- p1(X1), p2(X2), ..., pn(Xn) is the associated probability densities.
- x1, x2, ..., xn is a set of corresponding values for those random variables.

Then the maximum likelihood estimate is a value that maximizes the simultaneous probability of occurrence of `x1, x2, ..., xn`.

As an example, if each `Xi` is normally distributed with the mean `mu` and standard deviation sigma, then the value that maximizes the product of the probabilities `p(x1), p(x2),..., p(xn)` is the mean of `{x1, x2, ..., xn}` as shown here:

```
                 n
theta = (1/n)*SUM xi
                i=1
```

Also keep in mind that the probabilities `p(x1), p(x2),..., p(xn)` will be functions that include the unknown mean and standard deviation. Consequently, maximizing the likelihood function yields an expression for the mean.

A Python Code Sample for MLE

Listing 6.2 displays the content of `mle.py` that shows you how to calculate the MLE for a set of randomly generated values.

Listing 6.2: mle.py

```
import numpy as np
from scipy.optimize import minimize
```

```
# Generate synthetic data from an exponential
distribution with λ=2:
np.random.seed(0)
lambda_true = 2
data = np.random.exponential(scale=1/lambda_true,
size=100)


# Defining the likelihood function
def neg_log_likelihood(params):
    lambda_ = params[0]
    likelihood = np.sum(np.log(lambda_) - lambda_ *
data)
    return -likelihood


# Finding the MLE using optimization
result = minimize(neg_log_likelihood, x0=[1],
bounds=[(0.01, 10)])
lambda_mle = result.x[0]


print(f'True lambda:   {lambda_true}')
print(f'MLE of lambda: {lambda_mle}')
```

Listing 6.2 starts with two `import` statements and then initializes the scalar variable `lambda_true` with the value 2. Next, the variable `data` is initialized with a set of synthetically generated set of values from an exponential distribution with a known λ value.

The next portion of Listing 6.2 defines the negative log-likelihood function `neg_log_likelihood()`. We want to minimize this function to find the MLE because maximizing the likelihood is equivalent to minimizing the negative log-likelihood.

The final portion of Listing 6.2 displays the values of `lambda_true` and the calculated value `lambda_mle`. Launch the code in Listing 6.2, and you will see the following output:

```
True lambda:   2
MLE of lambda: 2.177111445372909
```

If you want more details regarding MLEs, navigate to the following article:

https://en.wikipedia.org/wiki/Maximum_likelihood_estimation

SUMMARY

This chapter started with an introduction to hypothesis testing, the null hypothesis, and the alternate hypothesis. Next, you learned about the t-test and the z-test, along with a discussion about p-values and their relevance to a null hypothesis.

Then you learned about converting z-scores into their corresponding p-values. Finally, you learned about point estimation, confidence level, and confidence intervals.

INTRODUCTION TO PYTHON

This appendix contains an introduction to Python, with information about useful tools for installing Python modules, basic constructs, and how to work with some data types.

The first part of this appendix discusses how to install Python, some environment variables, and how to use the interpreter. You will see code samples and how to save code in text files that you launch from the command line. The second part of this appendix shows you how to work with simple data types, such as numbers, fractions, and strings. The final part of this appendix discusses exceptions and how to use them in scripts.

Note: The code samples in this book are for Python 3.x.

TOOLS FOR PYTHON

The Anaconda Python distribution is available for Windows, Linux, and Mac:

https://www.anaconda.com/download/

Anaconda is well-suited for modules such as NumPy and SciPy, and if you are a Windows user, Anaconda appears to be a better alternative.

easy_install and pip

Whenever you need to install a Python module, use either `easy_install` or `pip` with the following syntax:

```
easy_install <module-name>
pip install <module-name>
```

Note: Python-based modules are easy to install, whereas modules with code written in C are usually faster, but more difficult to manage in terms of installation.

virtualenv

The `virtualenv` tool enables you to create isolated Python environments:

https://virtualenv.pypa.io/en/latest/installation.html

`virtualenv` addresses the problem of preserving the correct dependencies and versions (and indirectly permissions) for different applications. (If you are a Python novice, you might not need `virtualenv` right now, but keep this tool in mind for later projects.)

IPython

Another useful tool is `IPython`:

http://ipython.org/install.html

Type `ipython` to invoke IPython from the command line:
```
ipython
```

The preceding command displays the following output:
```
Python 3.9.13 (main, May 24 2022, 21:28:12)
Type 'copyright', 'credits' or 'license' for more
information
IPython 8.14.0 -- An enhanced Interactive Python. Type
'?' for help.


In [1]:
```

Now type a question mark ("?") at the prompt and you will see some useful information, a portion of which is here:
```
IPython -- An enhanced Interactive Python
=========================================


IPython offers a fully compatible replacement for the
standard Python
interpreter, with convenient shell features, special
commands, command
```

```
history mechanism and output results caching.

At your system command line, type 'ipython -h' to see
the command line
options available. This document only describes
interactive features.

GETTING HELP
------------

Within IPython you have various way to access help:

  ?          -> Introduction and overview of IPython's
features (this screen).
  object?   -> Details about 'object'.
  object??  -> More detailed, verbose information about
'object'.
  %quickref -> Quick reference of all IPython specific
syntax and magics.
  help      -> Access Python's own help system.

If you are in terminal IPython you can quit this screen
by pressing `q`.
```

Finally, type `quit` at the command prompt, and you will exit the IPython shell.

The next section shows you how to check whether Python is installed on your machine, and also where you can download Python.

PYTHON INSTALLATION

Before you download anything, check if you have Python already installed on your machine (which is likely if you have a MacBook or a Linux machine) by typing the following command in a command shell:

```
python -V
```

The output for the MacBook used in this book is here:

```
Python 3.9.1
```

Note: Install Python 3.9 (or as close as possible to this version) on your machine so that you will have the same version of Python that was used to test the Python scripts in this book.

If you need to install Python on your machine, navigate to the Python home page and select the "downloads" link or navigate directly to this website:

http://www.python.org/download

In addition, `PythonWin` is available for Windows:

http://www.cgl.ucsf.edu/Outreach/pc204/pythonwin.html

Use any text editor that can create, edit, and save Python scripts and save them as plain text files (do not use Microsoft Word). After you have Python installed and configured on your machine, you are ready to work with the scripts in this book.

SETTING THE PATH ENVIRONMENT VARIABLE (WINDOWS ONLY)

The `PATH` environment variable specifies a list of directories that are searched whenever you specify an executable program from the command line. A good guide to setting up your environment so that the Python executable is always available in every command shell is to follow the instructions here:

http://www.blog.pythonlibrary.org/2011/11/24/python-101-setting-up-python-on-windows/

LAUNCHING PYTHON ON YOUR MACHINE

There are three different ways to launch Python:

- Use the Python interactive interpreter.
- Launch Python scripts from the command line.
- Use an integrated development environment (IDE).

The next section shows you how to launch the interpreter from the command line. Later, you will learn about IDEs and how to launch Python scripts from the command line.

Note: The emphasis in this book is on launching Python scripts from the command line or entering code into the interpreter.

The Python Interactive Interpreter

Launch the Python interactive interpreter from the command line by opening a command shell and typing the following command:

```
python
```

You will see the following prompt (or something similar):

```
Python 3.9.1 (v3.9.1:1e5d33e9b9, Dec  7 2020, 12:44:01)
[Clang 12.0.0 (clang-1200.0.32.27)] on darwin
Type "help", "copyright", "credits" or "license" for
more information.
>>>
```

Now, type the expression 2 + 7 at the prompt:

```
>>> 2 + 7
```

Python displays the following result:

```
9
>>>
```

Type quit() to exit the Python shell.
You can launch any Python script from the command line by preceding it with the word python. For example, if you have a script myscript.py that contains Python commands, launch the script as follows:

```
python myscript.py
```

As a simple illustration, suppose that the script myscript.py contains the following Python code:

```
print('Hello World from Python')
print('2 + 7 = ', 2+7)
```

When you launch the preceding Python script, you will see the following output:

```
Hello World from Python
2 + 7 =  9
```

IDENTIFIERS

An *identifier* is the name of a variable, function, class, module, or other Python object. A valid identifier conforms to the following rules:

- starts with a letter (A to Z or a to z) or an underscore (_)
- zero or more letters, underscores, and digits (0 to 9)

    Note: Python identifiers cannot contain characters such as @, $, and %. Python is a case-sensitive language, so Abc and abc are different identifiers. In addition, Python has the following naming conventions:

- Class names start with an uppercase letter and all other identifiers with a lowercase letter.
- An initial underscore is used for private identifiers.
- Two initial underscores are used for strongly private identifiers.

    An identifier with two initial underscore characters and two trailing underscore characters indicates a language-defined special name.

LINES, INDENTATION, AND MULTI-LINE STATEMENTS

Unlike other programming languages (such as Java or Objective-C), Python uses indentation instead of curly braces for code blocks. Indentation must be consistent in a code block, as shown here:

```
if True:
    print("ABC")
    print("DEF")
else:
    print("ABC")
    print("DEF")
```

Multi-line statements can terminate with a new line or the backslash ("\") character, as shown here:

```
total = x1 + \
        x2 + \
        x3
```

Obviously, you can place x1, x2, and x3 on the same line, so there is no reason to use three separate lines; however, this functionality is available if you need to add a set of variables that do not fit on a single line.

You can specify multiple statements in one line by using a semicolon (";") to separate each statement, as shown here:

```
a=10; b=5; print(a); print(a+b)
```

The output of the preceding code snippet is here:

```
10
15
```

Note: The use of semi-colons and the continuation character are discouraged in Python.

QUOTATION MARKS AND COMMENTS

Python allows single ('), double ("), and triple (''' or """) quotation marks for string literals, provided that they match at the beginning and the end of the string. You can use triple quotation marks for strings that span multiple lines. The following examples are legal Python strings:

```
word = 'word'
line = "This is a sentence."
para = """This is a paragraph. This paragraph contains
more than one sentence."""
```

A string literal that begins with the letter "r" (for "raw") treats everything as a literal character and "escapes" the meaning of meta characters, as shown here:

```
a1 = r'\n'
a2 = r'\r'
a3 = r'\t'
print('a1:',a1,'a2:',a2,'a3:',a3)
```

The output of the preceding code block is here:

```
a1: \n a2: \r a3: \t
```

You can embed a single quotation mark in a pair of double quotation marks (and vice versa) to display a single quotation mark or a double quotation mark. Another way to accomplish the same result is to precede single or double quotation marks with a backslash ("\") character. The following code block illustrates these techniques:

```
b1 = "'"
b2 = '"'
b3 = '\''
b4 = «\»»
print('b1:',b1,'b2:',b2)
print('b3:',b3,'b4:',b4)
```

The output of the preceding code block is here:

```
b1: ' b2: "
b3: ' b4: "
```

A hash sign (#) that is not inside a string literal is the character that indicates the beginning of a comment. Moreover, all characters after the # and up to the physical line end are part of the comment (and are ignored by the Python interpreter). Consider the following code block:

```
#!/usr/bin/python
# First comment
print("Hello, Python!")  # second comment
```

This will produce following result:

```
Hello, Python!
```

A comment may be on the same line after a statement or expression:

```
name = "Tom Jones" # This is also comment
```

You can comment multiple lines as follows:

```
# This is comment one
# This is comment two
# This is comment three
```

A blank line in Python is a line containing only whitespace, a comment, or both.

SAVING YOUR CODE IN A MODULE

Earlier, you saw how to launch the Python interpreter from the command line and then enter commands. However, everything that you type in the interpreter is only valid for the current session. If you exit the interpreter and then launch the interpreter again, your previous definitions are no longer valid. Fortunately, Python enables you to store code in a text file, as discussed in the next section.

A module in Python is a text file that contains Python statements. In the previous section, you saw how the interpreter enables you to test code snippets whose definitions are valid for the current session. If you want to retain the code snippets and other definitions, place them in a text file so that you can execute that code outside of the interpreter.

The statements in a program are executed from top to bottom when the module is imported for the first time, which will then set up its variables and functions.

A module can be run directly from the command line, as shown here:

```
python First.py
```

As an illustration, place the following two statements in a text file called `First.py`:

```
x = 3
print(x)
```

Now type the following command:

```
python First.py
```

The output from the preceding command is 3, which is the same as executing the preceding code from the Python interpreter.

When a Python module is run directly, the special variable __name__ is set to __main__. You will often see the following type of code in a Python module:

```
if __name__ == '__main__':
    # do something here
    print('Running directly')
```

The preceding code snippet enables Python to determine whether a Python module was launched from the command line or imported into another Python module.

SOME STANDARD MODULES

The Python Standard Library provides many modules that can simplify your own Python scripts. A list of the Standard Library modules is here:

http://www.python.org/doc

Some of the most important modules include `cgi`, `math`, `os`, `pickle`, `random`, `re`, `socket`, `sys`, `time`, and `urllib`.

The code samples in this book use the modules `math`, `os`, `random`, `re`, `socket`, `sys`, `time`, and `urllib`. You need to import these modules in order to use them in your code. For example, the following code block shows you how to import four standard Python modules:

```
import datetime
import re
import sys
import time
```

The code samples in this book import one or more of the preceding modules, as well as other Python modules.

THE HELP() AND DIR() FUNCTIONS

An Internet search for Python-related topics usually returns a number of links with useful information. Alternatively, you can check the official Python documentation site: *docs.python.org*.

In addition, Python provides the `help()` and `dir()` functions that are accessible from the interpreter. The `help()` function displays documentation strings, whereas the `dir()` function displays defined symbols. For example, if you type `help(sys)`, you will see documentation for the `sys` module, whereas `dir(sys)` displays a list of the defined symbols.

Type the following command in the interpreter to display the string-related methods:

```
>>> dir(str)
```

The preceding command generates the following output:

```
['__add__', '__class__', '__contains__', '__delattr__',
'__doc__', '__eq__', '__format__', '__ge__', '__
getattribute__', '__getitem__', '__getnewargs__',
```

```
'__getslice__', '__gt__', '__hash__', '__init__',
'__le__', '__len__', '__lt__', '__mod__', '__mul__',
'__ne__', '__new__', '__reduce__', '__reduce_ex__',
'__repr__', '__rmod__', '__rmul__', '__setattr__',
'__sizeof__', '__str__', '__subclasshook__', '_
formatter_field_name_split', '_formatter_parser',
'capitalize', 'center', 'count', 'decode', 'encode',
'endswith', 'expandtabs', 'find', 'format', 'index',
'isalnum', 'isalpha', 'isdigit', 'islower', 'isspace',
'istitle', 'isupper', 'join', 'ljust', 'lower',
'lstrip', 'partition', 'replace', 'rfind', 'rindex',
'rjust', 'rpartition', 'rsplit', 'rstrip', 'split',
'splitlines', 'startswith', 'strip', 'swapcase',
'title', 'translate', 'upper', 'zfill']
```

The preceding list gives you a consolidated inventory of built-in functions (including some that are discussed later in this appendix). Although the max() function obviously returns the maximum value of its arguments, the purpose of other functions, such as filter() or map(), is not immediately apparent (unless you have used them in other programming languages). The preceding list provides a starting point for finding out more about various Python built-in functions that are not discussed in this appendix.

Note that while dir() does not list the names of built-in functions and variables, you can obtain this information from the standard module __builtin__ that is automatically imported under the name __builtins__:

```
>>> dir(__builtins__)
```

The following command shows you how to get more information about a function:

```
help(str.lower)
```

The output from the preceding command is here:

```
Help on method_descriptor:


lower(...)
    S.lower() -> string


    Return a copy of the string S converted to
lowercase.
(END)
```

Check the online documentation and also experiment with `help()` and `dir()` when you need additional information about a particular function or module.

COMPILE TIME AND RUNTIME CODE CHECKING

Python performs some compile-time checking, but most checks (including type, name, and so forth) are *deferred* until code execution. Consequently, if your code references a user-defined function that does not exist, the code will compile successfully. In fact, the code will fail with an exception *only* when the code execution path references the non-existent function.

As a simple example, consider the following function `myFunc` that references the non-existent function called `DoesNotExist`:

```
def myFunc(x):
    if x == 3:
        print(DoesNotExist(x))
    else:
        print('x: ',x)
```

The preceding code will only fail when the `myFunc` function is passed the value 3, after which Python raises an error.

Uninitialized Variables and the Value None

Python distinguishes between an uninitialized variable and the value `None`. The former is a variable that has not been assigned a value, whereas the value `None` is a value that indicates "no value." Collections and methods often return the value `None`, and you can test for the value `None` in conditional logic.

Now that you understand some basic concepts (such as how to use the Python interpreter) and how to launch your custom Python modules, the next section discusses primitive data types in Python.

SIMPLE DATA TYPES

Python supports primitive data types, such as numbers (integers, floating point numbers, and exponential numbers), strings, and dates. Python also supports more complex data types, such as lists (or arrays), tuples, and dictionaries. The next several sections discuss some of the primitive data types, along with code snippets that show you how to perform various operations on those data types.

WORKING WITH NUMBERS

Python provides arithmetic operations for manipulating numbers in a straight-forward manner that is similar to other programming languages. The following examples involve arithmetic operations on integers:

```
>>> 2+2
4
>>> 4/3
1
>>> 3*8
24
```

The following example assigns numbers to two variables and computes their product:

```
>>> x = 4
>>> y = 7
>>> x * y
28
```

The following examples demonstrate arithmetic operations involving integers:

```
>>> 2+2
4
>>> 4/3
1
>>> 3*8
24
```

Notice that division ("/") of two integers is actually truncation in which only the integer result is retained. The following example converts a floating-point number into exponential form:

```
>>> fnum = 0.00012345689000007
>>> "%.14e"%fnum
```

```
'1.23456890000070e-04'
```

You can use the `int()` function and the `float()` function to convert strings to numbers:

```
word1 = "123"
word2 = "456.78"
var1 = int(word1)
var2 = float(word2)
print("var1: ",var1," var2: ",var2)
```

The output from the preceding code block is here:

```
var1:  123  var2:  456.78
```

Alternatively, you can use the `eval()` function:

```
word1 = "123"
word2 = "456.78"
var1 = eval(word1)
var2 = eval(word2)
print("var1: ",var1," var2: ",var2)
```

If you attempt to convert a string that is not a valid integer or a floating-point number, Python raises an exception, so it is advisable to place your code in a `try/except` block.

Working with Other Bases

Numbers in Python are in base 10 (the default), but you can easily convert numbers to other bases. For example, the following code block initializes the variable x with the value `1234`, and then displays that number in base 2, 8, and 16, respectively:

```
>>> x = 1234
>>> bin(x) '0b10011010010'
>>> oct(x) '0o2322'
>>> hex(x) '0x4d2'
```

Use the `format()` function if you want to suppress the 0b, 0o, or 0x prefixes, as shown here:

```
>>> format(x, 'b') '10011010010'
>>> format(x, 'o') '2322'
```

```
>>> format(x, 'x') '4d2'
```

Negative integers are displayed with a negative sign:

```
>>> x = -1234
>>> format(x, 'b') '-10011010010'
>>> format(x, 'x') '-4d2'
```

The chr() Function

The `chr()` function takes a positive integer as a parameter and converts it to its corresponding alphabetic value (if one exists). The letters A through Z have decimal representations of 65 through 91 (which correspond to hexadecimal 41 through 5b), and the lowercase letters a through z have decimal representations of 97 through 122 (hexadecimal 61 through 7b).

Here is an example of using the `chr()` function to print an uppercase A:

```
>>> x=chr(65)
>>> x
'A'
```

The following code block prints the ASCII values for a range of integers:

```
result = ""
for x in range(65,91):
  print(x, chr(x))
  result = result+chr(x)+' '
print("result: ",result)
```

Note: Python 2 uses ASCII strings whereas Python 3 uses UTF-8. You can represent a range of characters with the following line:

```
for x in range(65,91):
```

However, the following equivalent code snippet is more intuitive:

```
for x in range(ord('A'), ord('Z')):
```

If you want to display the result for lowercase letters, change the preceding range from `(65,91)` to either of the following statements:

```
for x in range(65,91):
for x in range(ord('a'), ord('z')):
```

The round() Function

The `round()` function enables you to round decimal values to the nearest precision:

```
>>> round(1.23, 1)
1.2
>>> round(-3.42,1)
-3.4
```

Formatting Numbers

Python allows you to specify the number of decimal places of precision to use when printing decimal numbers, as shown here:

```
>>> x = 1.23456
>>> format(x, '0.2f')
'1.23'
>>> format(x, '0.3f')
'1.235'
>>> 'value is {:0.3f}'.format(x) 'value is 1.235'
>>> from decimal import Decimal
>>> a = Decimal('4.2')
>>> b = Decimal('2.1')
>>> a + b
Decimal('6.3')
>>> print(a + b)
6.3
>>> (a + b) == Decimal('6.3')
True
>>> x = 1234.56789
>>> # Two decimal places of accuracy
>>> format(x, '0.2f')
'1234.57'
>>> # Right justified in 10 chars, one-digit accuracy
>>> format(x, '>10.1f')
'  1234.6'
>>> # Left justified
```

```
>>> format(x, '<10.1f') '1234.6 '
>>> # Centered
>>> format(x, '^10.1f') ' 1234.6 '
>>> # Inclusion of thousands separator
>>> format(x, ',')
'1,234.56789'
>>> format(x, '0,.1f')
'1,234.6'
```

WORKING WITH FRACTIONS

Python supports the `Fraction()` function (defined in the `fractions` module), which accepts two integers that represent the numerator and the denominator (which must be non-zero) of a fraction. Several example of defining and manipulating fractions in Python are shown here:

```
>>> from fractions import Fraction
>>> a = Fraction(5, 4)
>>> b = Fraction(7, 16)
>>> print(a + b)
27/16
>>> print(a * b) 35/64
>>> # Getting numerator/denominator
>>> c = a * b
>>> c.numerator
35
>>> c.denominator 64
>>> # Converting to a float >>> float(c)
0.546875
>>> # Limiting the denominator of a value
>>> print(c.limit_denominator(8))
4
>>> # Converting a float to a fraction >>> x = 3.75
>>> y = Fraction(*x.as_integer_ratio())
>>> y
Fraction(15, 4)
```

Before delving into code samples that work with strings, the next section briefly discusses Unicode and UTF-8, both of which are character encodings.

UNICODE AND UTF-8

A Unicode string consists of a sequence of numbers that are between 0 and 0x10ffff, where each number represents a group of bytes. An *encoding* is the manner in which a Unicode string is translated into a sequence of bytes. Among the various encodings, UTF-8 ("Unicode Transformation Format") is perhaps the most common, and it is also the default encoding for many systems. The digit 8 in UTF-8 indicates that the encoding uses 8-bit numbers, whereas UTF-16 uses 16-bit numbers (but this encoding is less common).

The ASCII character set is a subset of UTF-8, so a valid ASCII string can be read as a UTF-8 string without any re-encoding required. In addition, a Unicode string can be converted into a UTF-8 string.

Working with Unicode

Since Python supports Unicode, you can render characters in different languages. Unicode data can be stored and manipulated in the same way as strings. Create a Unicode string by prepending the letter u, as shown here:

```
>>> u'Hello from Python!'
u'Hello from Python!'
```

Special characters can be included in a string by specifying their Unicode value. For example, the following Unicode string embeds a space (which has the Unicode value 0x0020) in a string:

```
>>> u'Hello\u0020from Python!'
u'Hello from Python!'
```

Listing A.1 displays the content of Unicode1.py that illustrates how to display a string of characters in Japanese and another string of characters in Chinese (Mandarin).

Listing A.1: Unicode1.py

```
chinese1 = u'\u5c07\u63a2\u8a0e HTML5 \u53ca\u5176\
u4ed6'
hiragana = u'D3 \u306F \u304B\u3063\u3053\u3043\u3043 \
u3067\u3059!'
```

```
print('Chinese:',chinese1)
print('Hiragana:',hiragana)
```

The output of Listing A.1 is here:

```
Chinese: 將探討 HTML5 及其他
Hiragana: D3 は かっこいい です!
```

The next portion of this appendix shows you how to manage text strings with built-in Python functions.

WORKING WITH STRINGS

A string in Python 2 is a sequence of ASCII-encoded bytes, whereas a string in Python 3 is based on Unicode. You can concatenate two strings using the "+" operator. The following example prints a string and then concatenates two single-letter strings:

```
>>> 'abc'
'abc'
>>> 'a' + 'b'
'ab'
```

You can use "+" or "*" to concatenate identical strings, as shown here:

```
>>> 'a' + 'a' + 'a'
'aaa'
>>> 'a' * 3
'aaa'
```

You can assign strings to variables and print them using the print() statement:

```
>>> print('abc')
abc
>>> x = 'abc'
>>> print(x)
abc
>>> y = 'def'
>>> print(x + y)
abcdef
```

You can "unpack" the letters of a string and assign them to variables, as shown here:

```
>>> str = "World"
>>> x1,x2,x3,x4,x5 = str
>>> x1
'W'
>>> x2
'o'
>>> x3
'r'
>>> x4
'l'
>>> x5
'd'
```

The preceding code snippets shows you how easy it is to extract the letters in a text string. You can extract substrings of a string as shown in the following examples:

```
>>> x = "abcdef"
>>> x[0]
'a'
>>> x[-1]
'f'
>>> x[1:3]
'bc'
>>> x[0:2] + x[5:]
'abf'
```

However, you will cause an error if you attempt to subtract two strings, as you probably expect:

```
>>> 'a' - 'b'
Traceback (most recent call last):
  File "<stdin>", line 1, in <module>
TypeError: unsupported operand type(s) for -: 'str' and 'str'
```

The `try/except` construct allows you to handle the preceding type of exception so that the Python code can continue execution.

Comparing Strings

You can use the methods `lower()` and `upper()` to convert a string to lowercase and uppercase, respectively, as shown here:

```
>>> 'Python'.lower()
'python'
>>> 'Python'.upper()
'PYTHON'
>>>
```

The methods `lower()` and `upper()` are useful for performing a case-insensitive comparison of two ASCII strings. Listing A.2 displays the content of `Compare.py` that uses the `lower()` function to compare two ASCII strings.

Listing A.2: Compare.py

```
x = 'Abc'
y = 'abc'

if(x == y):
   print('x and y: identical')
elif (x.lower() == y.lower()):
   print('x and y: case insensitive match')
else:
   print('x and y: different')
```

Since x contains mixed-case letters and y contains lowercase letters, Listing A.2 displays the following output:

```
x and y: different
```

Formatting Strings

Python provides the functions `string.lstring()`, `string.rstring()`, and `string.center()` for positioning a text string so that it is left-justified, right-justified, and centered, respectively. As you saw in a previous section, Python also provides the `format()` method for advanced interpolation features.

Now enter the following commands in the interpreter:

```
import string

str1 = 'this is a string'
print(string.ljust(str1, 10))
print(string.rjust(str1, 40))
print(string.center(str1,40))
```

The output is shown here:

```
this is a string
                        this is a string
            this is a string
```

The next portion of this appendix shows you how to "slice and dice" text strings with built-in Python functions.

SLICING AND SPLICING STRINGS

Python enables you to extract substrings of a string (called "slicing") using array notation. Slice notation is start:stop:step, where the start, stop, and step values are integers that specify the start value, end value, and the increment value, respectively. The interesting part about slicing in Python is that you can use the value –1, which operates from the right-side instead of the left-side of a string.

Some examples of slicing a string are here:

```
text1 = "this is a string"
print('First 7 characters:',text1[0:7])
print('Characters 2-4:',text1[2:4])
print('Right-most character:',text1[-1])
print('Right-most 2 characters:',text1[-3:-1])
```

The output from the preceding code block is here:

```
First 7 characters: this is
Characters 2-4: is
Right-most character: g
Right-most 2 characters: in
```

Later in this appendix, you will see how to insert a string in the middle of another string.

Testing for Digits and Alphabetic Characters

Python enables you to examine each character in a string and then test whether that character is a digit or an alphabetic character. This section provides a simple introduction to regular expressions.

Listing A.3 displays the content of CharTypes.py that illustrates how to determine whether a string contains digits or characters. Although we have not discussed if statements, the examples in Listing A.3 are straightforward.

Listing A.3: CharTypes.py

```
str1 = "4"
str2 = "4234"
str3 = "b"
str4 = "abc"
str5 = "a1b2c3"

if(str1.isdigit()):
  print("this is a digit:",str1)

if(str2.isdigit()):
  print("this is a digit:",str2)

if(str3.isalpha()):
  print("this is alphabetic:",str3)

if(str4.isalpha()):
  print("this is alphabetic:",str4)

if(not str5.isalpha()):
  print("this is not pure alphabetic:",str5)

print("capitalized first letter:",str5.title())
```

Listing A.3 initializes some variables, followed by two conditional tests that check whether `str1` and `str2` are digits using the `isdigit()` function. The next portion of Listing A.3 checks if `str3`, `str4`, and `str5` are alphabetic strings using the `isalpha()` function. The output of Listing A.3 is here:

```
this is a digit: 4
this is a digit: 4234
this is alphabetic: b
this is alphabetic: abc
this is not pure alphabetic: a1b2c3
capitalized first letter: A1B2C3
```

SEARCH AND REPLACE A STRING IN OTHER STRINGS

Python provides methods for searching and replacing a string in a second text string. Listing A.4 displays the content of `FindPos1.py` that shows you how to use the find function to search for the occurrence of one string in another string.

Listing A.4: FindPos1.py

```
item1 = 'abc'
item2 = 'Abc'
text = 'This is a text string with abc'

pos1 = text.find(item1)
pos2 = text.find(item2)

print('pos1=',pos1)
print('pos2=',pos2)
```

Listing A.4 initializes the variables `item1`, `item2`, and `text`, and then searches for the index of the contents of `item1` and `item2` in the string text. The `find()` function returns the column number where the first successful match occurs; otherwise, the `find()` function returns a `-1` if a match is unsuccessful. The output from launching Listing A.4 is here:

```
pos1= 27
pos2= -1
```

In addition to the `find()` method, you can use the `in` operator when you want to test for the presence of an element, as shown here:

```
>>> lst = [1,2,3]
>>> 1 in lst
True
```

Listing A.5 displays the content of `Replace1.py` that shows you how to replace one string with another string.

Listing A.5: Replace1.py

```
text = 'This is a text string with abc'
print('text:',text)
text = text.replace('is a', 'was a')
print('text:',text)
```

Listing A.5 starts by initializing the variable text and then printing its contents. The next portion of Listing A.5 replaces the occurrence of "is a" with "was a" in the string text, and then prints the modified string. The output from launching Listing A.5 is here:

```
text: This is a text string with abc
text: This was a text string with abc
```

REMOVE LEADING AND TRAILING CHARACTERS

Python provides the functions `strip()`, `lstrip()`, and `rstrip()` to remove characters in a text string. Listing A.6 displays the content of `Remove1.py` that shows you how to search for a string.

Listing A.6 Remove1.py

```
text = '   leading and trailing white space   '
print('text1:','x',text,'y')

text = text.lstrip()
print('text2:','x',text,'y')
```

```
text = text.rstrip()
print('text3:','x',text,'y')
```

Listing A.6 starts by concatenating the letter x and the contents of the variable text and then printing the result. The second part of Listing A.6 removes the leading white spaces in the string text and then appends the result to the letter x. The third part of Listing A.6 removes the trailing white spaces in the string text (note that the leading white spaces have already been removed) and then appends the result to the letter x.

The output from launching Listing A.6 is here:

```
text1: x     leading and trailing white space     y
text2: x leading and trailing white space     y
text3: x leading and trailing white space y
```

If you want to remove extra white spaces inside a text string, use the replace() function, as discussed in the previous section. The following example illustrates how this can be accomplished, which also contains the re module for regular expressions:

```
import re
text = ,a      b'
a = text.replace(' ', '')
b = re.sub('\s+', ' ', text)

print(a)
print(b)
```

The result is here:

```
ab
a b
```

PRINTING TEXT WITHOUT NEW LINE CHARACTERS

If you need to suppress white space and new lines between objects' output with multiple print statements, you can use concatenation or the write() function.

The first technique is to concatenate the string representations of each object using the str() function prior to printing the result. For example, run the following statement in Python:

```
x = str(9)+str(0xff)+str(-3.1)
print('x: ',x)
```

The output is shown here:

```
x:  9255-3.1
```

The preceding line contains the concatenation of the numbers 9 and 255 (which is the decimal value of the hexadecimal number 0xff) and -3.1.

Incidentally, you can use the str() function with modules and user-defined classes. An example involving the Python built-in module sys is here:

```
>>> import sys
>>> print(str(sys))
<module 'sys' (built-in)>
```

The following code snippet illustrates how to use the write() function to display a string:

```
import sys
write = sys.stdout.write
write('123')
write('123456789')
```

The output is here:

```
1231234567899
```

TEXT ALIGNMENT

Python provides the methods ljust(), rjust(), and center() for aligning text. The ljust() and rjust() functions left-justify and right-justify a text string, respectively, whereas the center() function will center a string. An example is shown in the following code block:

```
text = 'Hello World'
text.ljust(20)
'Hello World '
```

```
>>> text.rjust(20)
' Hello World'
>>> text.center(20)
' Hello World '
```

You can use the `format()` function to align text. Use the `<`, `>`, or `^` characters, along with a desired width, to right justify, left justify, and center the text, respectively. The following examples illustrate how you can specify text justification:

```
>>> format(text, '>20')
'         Hello World'
>>>
>>> format(text, '<20')
'Hello World         '
>>>
>>> format(text, '^20')
'    Hello World     '
>>>
```

WORKING WITH DATES

Python provides a rich set of date-related functions that are documented here:

http://docs.python.org/2/library/datetime.html

Listing A.7 displays the content of the script `Datetime2.py` that shows various date-related values, such as the current date and time; the day of the week, month, and year; and the time in seconds since the epoch.

Listing A.7: Datetime2.py

```
import time
import datetime

print("Time in seconds since the epoch: %s" %time.
time())
print("Current date and time: " , datetime.datetime.
now())
```

```
print("Or like this: " ,datetime.datetime.now().
strftime("%y-%m-%d-%H-%M"))


print("Current year: ", datetime.date.today().
strftime("%Y"))
print("Month of year: ", datetime.date.today().
strftime("%B"))
print("Week number of the year: ", datetime.date.
today().strftime("%W"))
print("Weekday of the week: ", datetime.date.today().
strftime("%w"))
print("Day of year: ", datetime.date.today().
strftime("%j"))
print("Day of the month : ", datetime.date.today().
strftime("%d"))
print("Day of week: ", datetime.date.today().
strftime("%A"))
```

Listing A.8 displays the output generated by running
the code in Listing A.7.

Listing A.8: datetime2.out

```
Time in seconds since the epoch: 1375144195.66
Current date and time:  2013-07-29 17:29:55.664164
Or like this:  13-07-29-17-29
Current year:  2013
Month of year:  July
Week number of the year:  30
Weekday of the week:  1
Day of year:  210
Day of the month :  29
Day of week:  Monday
```

Python also enables you to perform arithmetic calculates with date-
related values, as shown in the following code block:

```
>>> from datetime import timedelta
>>> a = timedelta(days=2, hours=6)
```

```
>>> b = timedelta(hours=4.5)
>>> c = a + b
>>> c.days
2
>>> c.seconds
37800
>>> c.seconds / 3600
10.5
>>> c.total_seconds() / 3600
58.5
```

Converting Strings to Dates

Listing A.9 displays the content of `String2Date.py` that illustrates how to convert a string to a date, as well as how to calculate the difference between two dates.

Listing A.9: String2Date.py

```
from datetime import datetime

text = '2014-08-13'
y = datetime.strptime(text, '%Y-%m-%d')
z = datetime.now()
diff = z - y
print('Date difference:',diff)
```

The output from Listing A.9 is shown here:

```
Date difference: -210 days, 18:58:40.197130
```

EXCEPTION HANDLING

Unlike JavaScript, you cannot add a number and a string in Python. However, you can detect an illegal operation using the `try/except` construct in Python, which is similar to the `try/catch` construct in languages such as JavaScript and Java.

An example of a `try/except` block is here:

```
try:
    x = 4
    y = 'abc'
    z = x + y
except:
    print 'cannot add incompatible types:', x, y
```

When you run the preceding code, the `print()` statement in the `except` code block is executed because the variables x and y have incompatible types.

Earlier, you learned that subtracting two strings throws an exception:

```
>>> 'a' - 'b'
Traceback (most recent call last):
    File "<stdin>", line 1, in <module>
TypeError: unsupported operand type(s) for -: 'str' and
'str'
```

A simple way to handle this situation is to use a `try/except` block:

```
>>> try:
...   print('a' - 'b')
... except TypeError:
...   print('TypeError exception while trying to
subtract two strings')
... except:
...   print('Exception while trying to subtract two
strings')
...
```

The output from the preceding code block is here:

```
TypeError exception while trying to subtract two
strings
```

As you can see, the preceding code block specifies the finer-grained exception called `TypeError`, followed by a "generic" `except` code block to handle all other exceptions that might occur during the execution of your Python code. This style is similar to the exception handling in Java code.

Listing A.10 displays the content of `Exception1.py` that illustrates how to handle various types of exceptions.

Listing A.10: Exception1.py

```
import sys

try:
    f = open('myfile.txt')
    s = f.readline()
    i = int(s.strip())
except IOError as err:
    print("I/O error: {0}".format(err))
except ValueError:
    print("Could not convert data to an integer.")
except:
    print("Unexpected error:", sys.exc_info()[0])
    raise
```

Listing A.10 contains a `try` block followed by three `except` statements. If an error occurs in the `try` block, the first `except` statement is compared with the type of exception that occurred. If there is a match, then the subsequent `print()` statement is executed, and the program terminates. If not, a similar test is performed with the second `except` statement. If neither `except` statement matches the exception, the third `except` statement handles the exception, which involves printing a message and then "raising" an exception.

Note that you can also specify multiple exception types in a single statement, as shown here:

```
except (NameError, RuntimeError, TypeError):
    print('One of three error types occurred')
```

The preceding code block is more compact, but you do not know which of the three error types occurred. Python allows you to define custom exceptions, but this topic is beyond the scope of this book.

HANDLING USER INPUT

Python enables you to read user input from the command line via the `input()` function or the `raw_input()` function. Typically, you assign user input to a variable, which will contain all characters that users enter from the keyboard.

User input terminates when users press the <return> key (which is included with the input characters). Listing A.11 displays the content of UserInput1. py that prompts users for their name and then uses interpolation to display a response.

Listing A.11: UserInput1.py

```
userInput = input("Enter your name: ")
print ("Hello %s, my name is Python" % userInput)
```

The output of Listing A.11 is here (assume that the user entered the word Dave):

```
Hello Dave, my name is Python
```

The print() statement in Listing A.11 uses string interpolation via %s, which substitutes the value of the variable after the % symbol. This functionality is obviously useful when you want to specify something that is determined at runtime.

User input can cause exceptions (depending on the operations that your code performs), so it is important to include exception-handling code.

Listing A.12 displays the content of UserInput2.py that prompts users for a string and attempts to convert the string to a number in a try/except block.

Listing A.12: UserInput2.py

```
userInput = input("Enter something: ")

try:
  x = 0 + eval(userInput)
  print('you entered the number:',userInput)
except:
  print(userInput,'is a string')
```

Listing A.12 adds the number 0 to the result of converting a user's input to a number. If the conversion was successful, a message with the user's input is displayed. If the conversion failed, the except code block consists of a print() statement that displays a message.

Note: This code sample uses the eval() function, which should be avoided so that your code does not evaluate arbitrary (and possibly destructive) commands.

Listing A.13 displays the content of `UserInput3.py` that prompts users for two numbers and attempts to compute their sum in a pair of `try/except` blocks.

Listing A.13: UserInput3.py

```
sum = 0

msg = 'Enter a number:'
val1 = input(msg)

try:
  sum = sum + eval(val1)
except:
  print(val1,'is a string')

msg = 'Enter a number:'
val2 = input(msg)

try:
  sum = sum + eval(val2)
except:
  print(val2,'is a string')

print('The sum of',val1,'and',val2,'is',sum)
```

Listing A.13 contains two `try` blocks, each of which is followed by an `except` statement. The first `try` block attempts to add the first user-supplied number to the variable `sum`, and the second `try` block attempts to add the second user-supplied number to the previously entered number. An error message occurs if either input string is not a valid number; if both are valid numbers, a message is displayed containing the input numbers and their sum. (Be sure to read the caveat regarding the `eval()` function that is mentioned earlier.)

PYTHON AND EMOJIS (OPTIONAL)

Listing A.14 displays the content `remove_emojis.py` that illustrates how to remove emojis from a text string.

Listing A.14: remove_emojis.py

```
import re
import emoji


text = "I want a Chicago deep dish pizza tonight \
U0001f600"
print("text:")
print(text)
print()


emoji_pattern = re.compile("[" "\U0001F1E0-\U0001F6FF"
"]+", flags=re.UNICODE)
text = emoji_pattern.sub(r"", text)
text = "".join([x for x in text if x not in emoji.
UNICODE_EMOJI])
print("text:")
print(text)
print()
```

Listing A.14 starts with two `import` statements, followed by initializing the variable `text` with a text string, whose contents are displayed. The next portion of Listing A.14 defines the variable `emoji_pattern` as a regular expression that represents a range of Unicode values for emojis.

The next portion of Listing A.14 sets the variable `text` equal to the set of non-emoji characters contained in the previously initialized value for `text` and then displays its contents. Launch the code in Listing A.14, and you will see the following output:

```
text:
I want a Chicago deep dish pizza tonight ☺


text:
I want a Chicago deep dish pizza tonight
```

COMMAND-LINE ARGUMENTS

Python provides a `getopt` module to parse command-line options and arguments, and the `sys` module gives access to any command-line arguments via the `sys.argv`. This serves two purposes:

- `sys.argv` is the list of command-line arguments.
- `len(sys.argv)` is the number of command-line arguments.

Here, `sys.argv[0]` is the program name, so if the program is called `test.py`, it matches the value of `sys.argv[0]`.

Now you can provide input values for a Python program on the command line instead of providing input values by prompting users for their input. As an example, consider the script `test.py` shown here:

```
#!/usr/bin/python
import sys
print('Number of arguments:',len(sys.argv),'arguments')
print('Argument List:', str(sys.argv))
```

Now run above script as follows:

```
python test.py arg1 arg2 arg3
```

This will produce following result:

```
Number of arguments: 4 arguments.
Argument List: ['test.py', 'arg1', 'arg2', 'arg3']
```

The ability to specify input values from the command line provides useful functionality. For example, suppose that you have a custom Python class that contains the methods `add` and `subtract` to add and subtract a pair of numbers.

You can use command-line arguments to specify which method to execute on a pair of numbers, as shown here:

```
python MyClass add 3 5
python MyClass subtract 3 5
```

This functionality is useful because you can programmatically execute different methods in a Python class, which means that you can write unit tests for your code as well.

Listing A.15 displays the content of `Hello.py` that shows you how to use `sys.argv` to check the number of command line parameters.

Listing A.15: Hello.py

```python
import sys

def main():
  if len(sys.argv) >= 2:
    name = sys.argv[1]
  else:
    name = 'World'
  print('Hello', name)

# Standard boilerplate to invoke the main() function
if __name__ == '__main__':
  main()
```

Listing A.15 defines the `main()` function that checks the number of command-line parameters: if this value is at least 2, then the variable `name` is assigned the value of the second parameter (the first parameter is `Hello.py`), otherwise `name` is assigned the value `Hello`. The `print()` statement then prints the value of the variable `name`.

The final portion of Listing A.15 uses conditional logic to determine whether to execute the `main()` function.

SUMMARY

This appendix started with an explanation of how to install Python, some environment variables, and how to use the interpreter. Next, you learned how to save code in text files that you launch from the command line.

In addition, you saw how to work with simple data types, such as numbers, fractions, and strings. Then you learned how to handle exceptions that can occur and how to use them in Python modules.

INTRODUCTION TO PANDAS

This appendix introduces you to Pandas and provides code samples that illustrate some of its useful features. If you are familiar with these topics, you may wish to skim through the material and code samples.

The first part of this appendix contains a brief introduction to Pandas. This section contains code samples that illustrate some features of data frames and a brief discussion of series, which are two of the main features of Pandas.

The second part of this appendix discusses various types of data frames that you can create, such as numeric and Boolean data frames. In addition, we discuss examples of creating data frames with NumPy functions and random numbers.

Note: Several code samples in this appendix reference the NumPy library for working with arrays and generating random numbers, which you can learn from online articles.

WHAT IS PANDAS?

Pandas is a Python package that is compatible with other Python packages, such as NumPy and Matplotlib. Install Pandas by opening a command shell and invoking this command for Python 3.x:

```
pip3 install pandas
```

In many ways, the semantics of the APIs in the Pandas library are similar to a spreadsheet, along with support for XSL, XML, HTML, and CSV file types. Pandas provides a data type called a *data frame* (similar to a Python dictionary) with an extremely powerful functionality.

```
Pandas data frames support a variety of input types,
such as ndarray, list, dict, or series.
```

The data type `series` is another mechanism for managing data. In addition to performing an online search for more details regarding `Series`, the following article contains a good introduction:

https://towardsdatascience.com/20-examples-to-master-pandas-series-bc4c68200324

Options and Settings

You can change the default values of environment variables, an example of which is shown here:

```
import pandas as pd

display_settings = {
    'max_columns': 8,
    'expand_frame_repr': True,  # Wrap to multiple
pages
    'max_rows': 20,
    'precision': 3,
    'show_dimensions': True
}


for op, value in display_settings.items():
pd.set_option("display.{}".format(op), value)
```

Include the preceding code block in your own code if you want Pandas to display a maximum of 20 rows and 8 columns, and floating-point numbers displayed with 3 decimal places. Set `expand_frame_rep` to `True` if you want the output to "wrap around" to multiple pages. The preceding `for` loop iterates through `display_settings` and sets the options equal to their corresponding values.

In addition, the following code snippet displays all Pandas options and their current values in your code:

```
print(pd.describe_option())
```

There are various other operations that you can perform with options and their values (such as the `pd.reset()` method for resetting values), as described in the Pandas user guide:

https://pandas.pydata.org/pandas-docs/stable/user_guide/options.html

Data Frames

In simplified terms, a Pandas data frame is a two-dimensional data structure, and it is convenient to think of the data structure in terms of rows and columns. Data frames can be labeled (rows as well as columns), and the columns can contain different data types. The source of the dataset for a data frame can be a data file, a database table, and a Web service. The data frame features include

- data frame methods
- data frame statistics
- grouping, pivoting, and reshaping
- handling missing data
- joining data frames

The code samples in this appendix show you almost all the features in the preceding list.

Data Frames and Data Cleaning Tasks

The specific tasks that you need to perform depend on the structure and contents of a dataset. In general, you will perform a workflow with the following steps, not necessarily always in this order (and some might be optional). All of the following steps can be performed with a Pandas data frame:

- read data into a data frame
- display top of a data frame
- display column data types
- display missing values
- replace NA with a value
- iterate through the columns
- provide statistics for each column
- find missing values
- total missing values
- percentage of missing values
- sort table values

- print summary information
- identify columns with > 50% of the data missing
- rename columns

This appendix contains sections that illustrate how to perform many of the steps in the preceding list.

Alternatives to Pandas

There are alternatives to Pandas that offer useful features, some of which are shown below:

- PySpark (for large datasets)
- Dask (for distributed processing)
- Modin (faster performance)
- Datatable (R data.table for Python)

The inclusion of these alternatives is not intended to diminish Pandas. Indeed, you might not need any of the functionality in the preceding list. However, you might need such functionality in the future, so it is worthwhile for you to know about these alternatives now (and there may be even more powerful alternatives at some point in the future).

A PANDAS DATA FRAME WITH A NUMPY EXAMPLE

Listing B.1 shows the content of pandas_df.py that illustrates how to define several data frames and display their contents.

Listing B.1: pandas_df.py

```
import pandas as pd
import numpy as np

myvector1 = np.array([1,2,3,4,5])
print("myvector1:")
print(myvector1)
print()
```

```
mydf1 = pd.Data frame(myvector1)
print("mydf1:")
print(mydf1)
print()


myvector2 = np.array([i for i in range(1,6)])
print("myvector2:")
print(myvector2)
print()


mydf2 = pd.Data frame(myvector2)
print("mydf2:")
print(mydf2)
print()


myarray = np.array([[10,30,20],
[50,40,60],[1000,2000,3000]])
print("myarray:")
print(myarray)
print()


mydf3 = pd.Data frame(myarray)
print("mydf3:")
print(mydf3)
print()
```

Listing B.1 starts with standard `import` statements for Pandas and NumPy, followed by the definition of two one-dimensional NumPy arrays and a two-dimensional NumPy array. Each NumPy variable is followed by a corresponding Pandas data frame (`mydf1`, `mydf2`, and `mydf3`). Launch the code in Listing B.1 to see the following output, and you can compare the NumPy arrays with the Pandas data frames:

```
myvector1:
[1 2 3 4 5]
```

```
mydf1:
   0
0  1
1  2
2  3
3  4
4  5

myvector2:
[1 2 3 4 5]

mydf2:
   0
0  1
1  2
2  3
3  4
4  5

myarray:
[[   10   30   20]
 [   50   40   60]
 [1000 2000 3000]]

mydf3:
      0     1     2
0    10    30    20
1    50    40    60
2  1000  2000  3000
```

By contrast, the following code block illustrates how to define two Pandas `Series` that are part of the definition of a Pandas data frame:

```
names = pd.Series(['SF', 'San Jose', 'Sacramento'])
sizes = pd.Series([852469, 1015785, 485199])
df = pd.Data frame({ 'Cities': names, 'Size': sizes })
```

```
print(df)
```

Create a Python file with the preceding code (along with the required `import` statement), and when you launch that code, you will see the following output:

```
     City name     sizes
0           SF    852469
1     San Jose   1015785
2    Sacramento    485199
```

DESCRIBING A PANDAS DATA FRAME

Listing B.2 shows the content of `pandas_df_describe.py`, which illustrates how to define a Pandas data frame that contains a 3x3 NumPy array of integer values, where the rows and columns of the data frame are labeled. Other aspects of the data frame are also displayed.

Listing B.2: pandas_df_describe.py

```
import numpy as np
import pandas as pd

myarray = np.array([[10,30,20],
[50,40,60],[1000,2000,3000]])

rownames = ['apples', 'oranges', 'beer']
colnames = ['January', 'February', 'March']

mydf = pd.Data frame(myarray, index=rownames,
columns=colnames)
print("contents of df:")
print(mydf)
print()

print("contents of January:")
print(mydf['January'])
print()
```

```
print("Number of Rows:")
print(mydf.shape[0])
print()

print("Number of Columns:")
print(mydf.shape[1])
print()

print("Number of Rows and Columns:")
print(mydf.shape)
print()

print("Column Names:")
print(mydf.columns)
print()

print("Column types:")
print(mydf.dtypes)
print()

print("Description:")
print(mydf.describe())
print()
```

Listing B.2 starts with two standard import statements followed by the variable myarray, which is a 3x3 NumPy array of numbers. The variables rownames and colnames provide names for the rows and columns, respectively, of the Pandas data frame mydf, which is initialized as a data frame with the specified data source (i.e., myarray).

The first portion of the following output requires a single print() statement (which simply displays the contents of mydf). The second portion of the output is generated by invoking the describe() method that is available for any Pandas data frame. The describe() method is useful: you will see various statistical quantities, such as the mean, standard deviation minimum, and maximum performed by *columns* (not rows), along with values for the 25th, 50th, and 75th percentiles. The output of Listing B.2 is here:

```
contents of df:
         January   February   March
apples        10         30      20
oranges       50         40      60
beer        1000       2000    3000


contents of January:
apples        10
oranges       50
beer        1000
Name: January, dtype: int64


Number of Rows:
3


Number of Columns:
3


Number of Rows and Columns:
(3, 3)


Column Names:
Index(['January', 'February', 'March'], dtype='object')


Column types:
January      int64
February     int64
March        int64
dtype: object


Description:
            January      February        March
count      3.000000      3.000000      3.000000
mean     353.333333    690.000000   1026.666667
```

std	560.386771	1134.504297	1709.073823
min	10.000000	30.000000	20.000000
25%	30.000000	35.000000	40.000000
50%	50.000000	40.000000	60.000000
75%	525.000000	1020.000000	1530.000000
max	1000.000000	2000.000000	3000.000000

BOOLEAN DATA FRAMES

Pandas supports Boolean operations on data frames, such as the logical OR, the logical AND, and the logical negation of a pair of data frames. Listing B.3 shows the content of pandas_boolean_df.py that illustrates how to define a data frame whose rows and columns are Boolean values.

Listing B.3: pandas_boolean_df.py

```
import pandas as pd

df1 = pd.Data frame({'a': [1, 0, 1], 'b': [0, 1, 1] },
dtype=bool)
df2 = pd.Data frame({'a': [0, 1, 1], 'b': [1, 1, 0] },
dtype=bool)

print("df1 & df2:")
print(df1 & df2)

print("df1 | df2:")
print(df1 | df2)

print("df1 ^ df2:")
print(df1 ^ df2)
```

Listing B.3 initializes the data frames df1 and df2, and then computes df1 & df2, df1 | df2, and df1 ^ df2, which represent the logical AND, the logical OR, and the logical negation, respectively, of df1 and df2. The output from launching the code in Listing B.3 is as follows:

```
df1 & df2:
       a       b
0   False   False
1   False    True
2    True   False
df1 | df2:
       a       b
0    True    True
1    True    True
2    True    True
df1 ^ df2:
       a       b
0    True    True
1    True   False
2   False    True
```

Transposing a Data Frame

The T attribute (as well as the transpose function) enables you to generate the transpose of a Pandas data frame, similar to the NumPy ndarray. The transpose operation switches rows to columns and columns to rows. For example, the following code snippet defines a data frame df1 and then displays the transpose of df1:

```
df1 = pd.Data frame({'a': [1, 0, 1], 'b': [0, 1, 1] },
dtype=int)

print("df1.T:")
print(df1.T)
```

The output of the preceding code snippet is here:

```
df1.T:
   0  1  2
a  1  0  1
b  0  1  1
```

The following code snippet defines data frames `df1` and `df2` and then displays their sum:

```
df1 = pd.Data frame({'a' : [1, 0, 1], 'b' : [0, 1, 1]
}, dtype=int)
df2 = pd.Data frame({'a' : [3, 3, 3], 'b' : [5, 5, 5]
}, dtype=int)


print("df1 + df2:")
print(df1 + df2)
```

The output is here:

```
df1 + df2:
   a  b
0  4  5
1  3  6
2  4  6
```

DATA FRAMES AND RANDOM NUMBERS

Listing B.4 shows the content of `pandas_random_df.py` that illustrates how to create a Pandas data frame with random integers.

Listing B.4: pandas_random_df.py

```
import pandas as pd
import numpy as np


df = pd.Data frame(np.random.randint(1, 5, size=(5,
2)), columns=['a','b'])
df = df.append(df.agg(['sum', 'mean']))


print("Contents of data frame:")
print(df)
```

Listing B.4 defines the data frame `df` that consists of 5 rows and 2 columns of random integers between 1 and 5. Notice that the columns of `df` are labeled "a" and "b." In addition, the next code snippet appends two rows

consisting of the sum and the mean of the numbers in both columns. The
output of Listing B.4 is here:

```
      a     b
0         1.0   2.0
1         1.0   1.0
2         4.0   3.0
3         3.0   1.0
4         1.0   2.0
sum    10.0   9.0
mean    2.0   1.8
```

Listing B.5 shows the content of `pandas_combine_df.py` that illustrates how to combine Pandas data frames.

Listing B.5: pandas_combine_df.py

```
import pandas as pd
import numpy as np

df = pd.Data frame({'foo1' : np.random.randn(5),
                    'foo2' : np.random.randn(5)})

print("contents of df:")
print(df)

print("contents of foo1:")
print(df.foo1)

print("contents of foo2:")
print(df.foo2)
```

Listing B.5 defines the data frame `df` that consists of 5 rows and 2 columns (labeled "`foo1`" and "`foo2`") of random real numbers between 0 and
5. The next portion of Listing B.5 shows the content of `df` and `foo1`. The
output of Listing B.5 is as follows:

```
contents of df:
        foo1      foo2
```

```
0   0.274680  _0.848669
1  _0.399771  _0.814679
2   0.454443  _0.363392
3   0.473753   0.550849
4  _0.211783  _0.015014
contents of foo1:
0      0.256773
1      1.204322
2      1.040515
3     _0.518414
4      0.634141
Name: foo1, dtype: float64
contents of foo2:
0     _2.506550
1     _0.896516
2     _0.222923
3      0.934574
4      0.527033
Name: foo2, dtype: float64
```

READING CSV FILES IN PANDAS

Pandas provides the `read_csv()` method for reading the contents of CSV files. For example, Listing B.6 shows the contents of `sometext.csv` that contain labeled data (`spam` or `ham`), and Listing B.7 shows the content of `read_csv_file.py` that illustrates how to read the contents of a CSV file.

Listing B.6: sometext.csv

```
type    text
ham     Available only for today
ham     I'm joking with you
spam    Free entry in 2 a wkly comp
ham     U dun say so early hor
ham     I don't think he goes to usf
```

```
spam    FreeMsg Hey there
ham     my brother is not sick
ham     As per your request Melle
spam    WINNER!! As a valued customer
```

Listing B.7: read_csv_file.py

```
import pandas as pd
import numpy as np


df = pd.read_csv('sometext.csv', delimiter='\t')


print("=> First five rows:")
print(df.head(5))
```

Listing B.7 reads the contents of `sometext.csv`, whose columns are separated by a tab ("\t") delimiter. Launch the code in Listing B.7 to see the following output:

```
=> First five rows:
    type                        text
0   ham     Available only for today
1   ham            I'm joking with you
2   spam  Free entry in 2 a wkly comp
3   ham          U dun say so early hor
4   ham   I don't think he goes to usf
```

The default value for the `head()` method is 5, but you can display the first n rows of a data frame `df` with the code snippet `df.head(n)`.

Specifying a Separator and Column Sets in Text Files

The previous section showed you how to use the `delimiter` attribute to specify the delimiter in a text file. You can also use the `sep` parameter to specify a different separator. In addition, you can assign the `names` parameter the column names in the data that you want to read. An example of using `delimiter` and `sep` is here:

```
df2 = pd.read_csv("data.csv",sep="|",
                  names=["Name","Surname","Height","We
ight"])
```

Pandas also provides the `read_table()` method for reading the contents of CSV files, which uses the same syntax as the `read_csv()` method.

Specifying an Index in Text Files

Suppose that you know that a particular column in a text file contains the index value for the rows in the text file. For example, a text file that contains the data in a relational table would typically contain an index column.

Fortunately, Pandas allows you to specify the kth column as the index in a text file, as shown here:

```
df = pd.read_csv('myfile.csv', index_col=k)
```

THE LOC() AND ILOC() METHODS

If you want to display the contents of a record in a Pandas data frame, specify the index of the row in the `loc()` method. For example, the following code snippet displays the data by feature name in a data frame `df`:

```
df.loc[feature_name]
```

Select the first row of the "height" column in the data frame:

```
df.loc([0], ['height'])
```

The following code snippet uses the `iloc()` function to display the first 8 records of the name column with this code snippet:

```
df.iloc[0:8]['name']
```

CONVERTING CATEGORICAL DATA TO NUMERIC DATA

One common task (especially in machine learning) involves converting a feature containing character data into a feature that contains numeric data. Listing B.8 shows the content of `cat2numeric.py` that illustrates how to replace a text field with a corresponding numeric field.

Listing B.8: cat2numeric.py

```
import pandas as pd
import numpy as np
```

```
df = pd.read_csv('sometext.csv', delimiter='\t')

print("=> First five rows (before):")
print(df.head(5))
print("-------------------------")
print()


# map ham/spam to 0/1 values:
df['type'] = df['type'].map( {'ham':0 , 'spam':1} )

print("=> First five rows (after):")
print(df.head(5))
print("-------------------------")
```

Listing B.8 initializes the data frame df with the contents of the CSV file sometext.csv, and then displays the contents of the first five rows by invoking df.head(5), which is also the default number of rows to display.

The next code snippet in Listing B.8 invokes the map() method to replace occurrences of ham with 0 and replace occurrences of spam with 1 in the column labeled type, as shown here:

```
df['type'] = df['type'].map( {'ham':0 , 'spam':1} )
```

The last portion of Listing B.8 invokes the head() method again to display the first five rows of the dataset after having renamed the contents of the column type. Launch the code in Listing B.8 to see the following output:

```
=> First five rows (before):
   type                         text
0  ham      Available only for today
1  ham              I'm joking with you
2  spam  Free entry in 2 a wkly comp
3  ham          U dun say so early hor
4  ham   I don't think he goes to usf
-------------------------

=> First five rows (after):
   type                         text
```

```
0     0       Available only for today
1     0                I'm joking with you
2     1   Free entry in 2 a wkly comp
3     0            U dun say so early hor
4     0   I don't think he goes to usf


-------------------------
```

As another example, Listing B.9 shows the content of shirts.csv and Listing B.10 shows the content of shirts.py; these examples illustrate four techniques for converting categorical data into numeric data.

Listing B.9: shirts.csv

```
type,ssize
shirt,xxlarge
shirt,xxlarge
shirt,xlarge
shirt,xlarge
shirt,xlarge
shirt,large
shirt,medium
shirt,small
shirt,small
shirt,xsmall
shirt,xsmall
shirt,xsmall
```

Listing B.10: shirts.py

```python
import pandas as pd

shirts = pd.read_csv("shirts.csv")
print("shirts before:")
print(shirts)
print()
```

```
# TECHNIQUE #1:
#shirts.loc[shirts['ssize']=='xxlarge','size'] = 4
#shirts.loc[shirts['ssize']=='xlarge', 'size'] = 4
#shirts.loc[shirts['ssize']=='large',  'size'] = 3
#shirts.loc[shirts['ssize']=='medium', 'size'] = 2
#shirts.loc[shirts['ssize']=='small',  'size'] = 1
#shirts.loc[shirts['ssize']=='xsmall', 'size'] = 1

# TECHNIQUE #2:
#shirts['ssize'].replace('xxlarge', 4, inplace=True)
#shirts['ssize'].replace('xlarge',  4, inplace=True)
#shirts['ssize'].replace('large',   3, inplace=True)
#shirts['ssize'].replace('medium',  2, inplace=True)
#shirts['ssize'].replace('small',   1, inplace=True)
#shirts['ssize'].replace('xsmall',  1, inplace=True)

# TECHNIQUE #3:
#shirts['ssize'] = shirts['ssize'].apply({'xxlarge':4,
'xlarge':4, 'large':3, 'medium':2, 'small':1,
'xsmall':1}.get)

# TECHNIQUE #4:
shirts['ssize'] = shirts['ssize'].
replace(regex='xlarge', value=4)
shirts['ssize'] = shirts['ssize'].
replace(regex='large',  value=3)
shirts['ssize'] = shirts['ssize'].
replace(regex='medium', value=2)
shirts['ssize'] = shirts['ssize'].
replace(regex='small',  value=1)

print("shirts after:")
print(shirts)
```

Listing B.10 starts with a code block of six statements that uses direct comparison with strings to make numeric replacements. For example, the following code snippet replaces all occurrences of the string `xxlarge` with the value 4:

```
shirts.loc[shirts['ssize']=='xxlarge','size'] = 4
```

The second code block consists of six statements that use the `replace()` method to perform the same updates, an example of which is shown here:

```
shirts['ssize'].replace('xxlarge', 4, inplace=True)
```

The third code block consists of a single statement that uses the `apply()` method to perform the same updates, as shown here:

```
shirts['ssize'] = shirts['ssize'].apply({'xxlarge':4,
'xlarge':4, 'large':3, 'medium':2, 'small':1,
'xsmall':1}.get)
```

The fourth code block consists of four statements that use regular expressions to perform the same updates, an example of which is shown here:

```
shirts['ssize'] = shirts['ssize'].
replace(regex='xlarge', value=4)
```

Since the preceding code snippet matches `xxlarge` as well as `xlarge`, we only need *four* statements instead of six statements. (If you are unfamiliar with regular expressions, you can find information online.) Launch the code in Listing B.10 to see the following output:

```
shirts before
       type     size
0     shirt   xxlarge
1     shirt   xxlarge
2     shirt    xlarge
3     shirt    xlarge
4     shirt    xlarge
5     shirt     large
6     shirt    medium
7     shirt     small
8     shirt     small
9     shirt    xsmall
10    shirt    xsmall
11    shirt    xsmall
```

```
shirts after:
        type   size
0      shirt       4
1      shirt       4
2      shirt       4
3      shirt       4
4      shirt       4
5      shirt       3
6      shirt       2
7      shirt       1
8      shirt       1
9      shirt       1
10     shirt       1
11     shirt       1
```

MATCHING AND SPLITTING STRINGS

Listing B.11 shows the content of `shirts_str.py`, which illustrates how to match a column value with an initial string and how to split a column value based on a letter.

Listing B.11: shirts_str.py

```
import pandas as pd

shirts = pd.read_csv("shirts2.csv")
print("shirts:")
print(shirts)
print()

print("shirts starting with xl:")
print(shirts[shirts.ssize.str.startswith('xl')])
print()
```

```
print("Exclude 'xlarge' shirts:")
print(shirts[shirts['ssize'] != 'xlarge'])
print()

print("first three letters:")
shirts['sub1'] = shirts['ssize'].str[:3]
print(shirts)
print()

print("split ssize on letter 'a':")
shirts['sub2'] = shirts['ssize'].str.split('a')
print(shirts)
print()

print("Rows 3 through 5 and column 2:")
print(shirts.iloc[2:5, 2])
print()
```

Listing B.11 initializes the data frame df with the contents of the CSV file shirts.csv, and then displays the contents of df. The next code snippet in Listing B.11 uses the startswith() method to match the shirt types that start with the letters xl, followed by a code snippet that displays the shorts whose size does not equal the string xlarge.

The next code snippet uses the construct str[:3] to display the first three letters of the shirt types, followed by a code snippet that uses the split() method to split the shirt types based on the letter "a."

The final code snippet invokes iloc[2:5,2] to display the contents of rows 3 through 5 inclusive, and only the second column. The output of Listing B.11 is as follows:

```
shirts:
      type     ssize
0    shirt   xxlarge
1    shirt   xxlarge
2    shirt    xlarge
3    shirt    xlarge
4    shirt    xlarge
```

```
5    shirt    large
6    shirt    medium
7    shirt     small
8    shirt     small
9    shirt   xsmall
10   shirt   xsmall
11   shirt   xsmall


shirts starting with xl:
     type    ssize
2   shirt   xlarge
3   shirt   xlarge
4   shirt   xlarge


Exclude 'xlarge' shirts:
      type     ssize
0    shirt   xxlarge
1    shirt   xxlarge
5    shirt     large
6    shirt    medium
7    shirt     small
8    shirt     small
9    shirt    xsmall
10   shirt    xsmall
11   shirt    xsmall


first three letters:
      type    ssize  sub1
0    shirt   xxlarge   xxl
1    shirt   xxlarge   xxl
2    shirt   xlarge    xla
3    shirt   xlarge    xla
4    shirt   xlarge    xla
5    shirt    large    lar
```

```
6    shirt    medium    med
7    shirt    small     sma
8    shirt    small     sma
9    shirt    xsmall    xsm
10   shirt    xsmall    xsm
11   shirt    xsmall    xsm

split ssize on letter 'a':
      type    ssize  sub1        sub2
0    shirt   xxlarge  xxl  [xxl, rge]
1    shirt   xxlarge  xxl  [xxl, rge]
2    shirt   xlarge   xla   [xl, rge]
3    shirt   xlarge   xla   [xl, rge]
4    shirt   xlarge   xla   [xl, rge]
5    shirt   large    lar    [l, rge]
6    shirt   medium   med   [medium]
7    shirt   small    sma   [sm, ll]
8    shirt   small    sma   [sm, ll]
9    shirt   xsmall   xsm  [xsm, ll]
10   shirt   xsmall   xsm  [xsm, ll]
11   shirt   xsmall   xsm  [xsm, ll]

Rows 3 through 5 and column 2:
2    xlarge
3    xlarge
4    xlarge
Name: ssize, dtype: object
```

CONVERTING STRINGS TO DATES

Listing B.12 shows the content of `string2date.py`, which illustrates how to convert strings to date formats.

Listing B.12: string2date.py

```
import pandas as pd

bdates1 = {'strdates':  ['20210413','20210813','202112
25'],
           'people': ['Sally','Steve','Sarah']
          }

df1 = pd.Data frame(bdates1, columns =
['strdates','people'])
df1['dates'] = pd.to_datetime(df1['strdates'],
format='%Y%m%d')
print("=> Contents of data frame df1:")
print(df1)
print()
print(df1.dtypes)
print()

bdates2 = {'strdates':  ['13Apr2021','08Aug2021','25D
ec2021'],
           'people': ['Sally','Steve','Sarah']
          }

df2 = pd.Data frame(bdates2, columns =
['strdates','people'])
df2['dates'] = pd.to_datetime(df2['strdates'],
format='%d%b%Y')
print("=> Contents of data frame df2:")
print(df2)
print()

print(df2.dtypes)
print()
```

Listing B.12 initializes the data frame df1 with the contents of bdates1, and then converts the strdates column to dates using the %Y%m%d format.

The next portion of Listing B.12 initializes the data frame df2 with the contents of bdates2, and then converts the strdates column to dates using the %d%b%Y format. Launch the code in Listing B.12 to see the following output:

```
=> Contents of data frame df1:
    strdates people      dates
0  20210413  Sally 2021-04-13
1  20210813  Steve 2021-08-13
2  20211225  Sarah 2021-12-25

strdates              object
people               object
dates        datetime64[ns]
dtype: object

=> Contents of data frame df2:
     strdates people      dates
0  13Apr2021  Sally 2021-04-13
1  08Aug2021  Steve 2021-08-08
2  25Dec2021  Sarah 2021-12-25

strdates              object
people               object
dates        datetime64[ns]
dtype: object
```

WORKING WITH DATE RANGES

Listing B.13 shows the content of pand_parse_dates.py that illustrates how to work with date ranges in a CSV file.

Listing B.13: pand_parse_dates.py

```
import pandas as pd

df = pd.read_csv('multiple_dates.csv', parse_
dates=['dates'])
```

```
print("df:")
print(df)
print()

df = df.set_index(['dates'])
start_d = "2021-04-30"
end_d   = "2021-08-31"

print("DATES BETWEEN",start_d,"AND",end_d,":")
print(df.loc[start_d:end_d])
print()

print("DATES BEFORE",start_d,":")
print(df.loc[df.index < start_d])

years = ['2020','2021','2022']
for year in years:
  year_sum = df.loc[year].sum()[0]
  print("SUM OF VALUES FOR YEAR",year,":",year_sum)
```

Listing B.13 starts by initializing the variable df with the contents of the CSV file multiple_dates.csv and then displaying its contents. The next code snippet sets the dates column as the index column and then initializes the variable start_d and end_d that contain a start date and an end date, respectively.

The next portion of Listing B.13 displays the dates between start_d and end_d, and then the list of dates that precede start_d. The final code block iterates through a list of years and then calculates the sum of the numbers in the values field for each year in the list. Launch the code in Listing B.13 to see the following output:

```
df:
        dates   values
0   2020-01-31    40.0
1   2020-02-28    45.0
2   2020-03-31    56.0
3   2021-04-30     NaN
```

```
4   2021-05-31      NaN
5   2021-06-30    140.0
6   2021-07-31     95.0
7   2022-08-31     40.0
8   2022-09-30     55.0
9   2022-10-31      NaN
10  2022-11-15     65.0


DATES BETWEEN 2021-04-30 AND 2021-08-31 :
              values
dates
2021-04-30      NaN
2021-05-31      NaN
2021-06-30    140.0
2021-07-31     95.0


DATES BEFORE 2021-04-30 :
              values
dates
2020-01-31     40.0
2020-02-28     45.0
2020-03-31     56.0


SUM OF VALUES FOR YEAR 2020 : 141.0
SUM OF VALUES FOR YEAR 2021 : 235.0
SUM OF VALUES FOR YEAR 2022 : 160.0
```

DETECTING MISSING DATES

Listing B.14 shows the content of `pandas_missing_dates.py` that illustrates how to detect missing date values in a CSV file.

Listing B.14: pandas_missing_dates.py

```
import pandas as pd

# A data frame from a dictionary of lists
data = {'Date': ['2021-01-18', '2021-01-20', '2021-01-
21', '2021-01-24'],
        'Name': ['Joe', 'John', 'Jane', 'Jim']}
df = pd.Data frame(data)

# Setting the Date values as index:
df = df.set_index('Date')

# to_datetime() converts string format to a DateTime
object:
df.index = pd.to_datetime(df.index)

start_d="2021-01-18"
end_d="2021-01-25"

# display dates that are not in the sequence:
print("MISSING DATES BETWEEN",start_d,"and",end_d,":")
dates = pd.date_range(start=start_d, end=end_d).
difference(df.index)

for date in dates:
  print("date:",date)
print()
```

Listing B.14 initializes the dictionary `data` with a list of values for the `Date` field and the `Name` field, after which the variable `df` is initialized as a data frame whose contents are from the `data` variable.

The next code snippet sets the `Date` field as the index of the data frame `df`, after which the string-based dates are converted to `DateTime` objects. Another pair of code snippets initialize the variable `start_d` and `end_d` with a start date and an end date, respectively.

The final portion of Listing B.14 initializes the variable `dates` with the list of missing dates between `start_d` and `end_d`, after which the contents of `dates` are displayed. Launch the code in Listing B.14 to see the following output:

```
MISSING DATES BETWEEN 2021-01-18 and 2021-01-25 :
date: 2022-01-19 00:00:00
date: 2022-01-22 00:00:00
date: 2022-01-23 00:00:00
date: 2022-01-25 00:00:00
```

INTERPOLATING MISSING DATES

Listing B.15 shows the contents of `missing_dates.csv`, and Listing B.16 shows the content of `pandas_interpolate.py` that illustrates how to replace `NaN` values with interpolated values that are calculated in several ways.

Listing B.15: missing_dates.csv

```
"dates","values"
2021-01-31,40
2021-02-28,45
2021-03-31,56
2021-04-30,NaN
2021-05-31,NaN
2021-06-30,140
2021-07-31,95
2021-08-31,40
2021-09-30,55
2021-10-31,NaN
2021-11-15,65
```

Notice the value 140 (shown in bold) in Listing B.15: this value is an outlier, which will affect the calculation of the interpolated values, and potentially generate additional outliers.

Listing B.16: pandas_interpolate.py

```python
import pandas as pd
df = pd.read_csv("missing_dates.csv")

# fill NaN values with linear interpolation:
df1 = df.interpolate()

# fill NaN values with quadratic polynomial
interpolation:
df2 = df.interpolate(method='polynomial', order=2)

# fill NaN values with cubic polynomial interpolation:
df3 = df.interpolate(method='polynomial', order=3)

print("original data frame:")
print(df)
print()
print("linear interpolation:")
print(df1)
print()
print("quadratic interpolation:")
print(df2)
print()
print("cubic interpolation:")
print(df3)
print()
```

Listing B.16 initializes df with the contents of the CSV file missing_ dates.csv and then initializes the three data frames df1, df2, and df3 that are based on linear, quadratic, and cubic interpolation, respectively, via the interpolate() method. Launch the code in Listing B.16 to see the following output:

```
original data frame:
        dates  values
0   2021-01-31    40.0
```

```
1    2021-02-28     45.0
2    2021-03-31     56.0
3    2021-04-30      NaN
4    2021-05-31      NaN
5    2021-06-30    140.0
6    2021-07-31     95.0
7    2021-08-31     40.0
8    2021-09-30     55.0
9    2021-10-31      NaN
10   2021-11-15     65.0


linear interpolation:
          dates   values
0    2021-01-31     40.0
1    2021-02-28     45.0
2    2021-03-31     56.0
3    2021-04-30     84.0
4    2021-05-31    112.0
5    2021-06-30    140.0
6    2021-07-31     95.0
7    2021-08-31     40.0
8    2021-09-30     55.0
9    2021-10-31     60.0
10   2021-11-15     65.0


quadratic interpolation:
          dates      values
0    2021-01-31   40.000000
1    2021-02-28   45.000000
2    2021-03-31   56.000000
3    2021-04-30   88.682998
4    2021-05-31  136.002883
5    2021-06-30  140.000000
6    2021-07-31   95.000000
```

```
7    2021-08-31    40.000000
8    2021-09-30    55.000000
9    2021-10-31    68.162292
10   2021-11-15    65.000000

cubic interpolation:
            dates       values
0     2021-01-31    40.000000
1     2021-02-28    45.000000
2     2021-03-31    56.000000
3     2021-04-30    92.748096
4     2021-05-31   132.055687
5     2021-06-30   140.000000
6     2021-07-31    95.000000
7     2021-08-31    40.000000
8     2021-09-30    55.000000
9     2021-10-31    91.479905
10    2021-11-15    65.000000
```

OTHER OPERATIONS WITH DATES

Listing B.17 shows the content of `pandas_misc1.py` for extracting a list of years from a column in a data frame.

Listing B.17: pandas_misc1.py

```
import pandas as pd
import numpy as np

df = pd.read_csv('multiple_dates.csv', parse_
dates=['dates'])
print("df:")
print(df)
print()
```

```
year_list = df['dates']

arr1 = np.array([])
for long_year in year_list:
  year = str(long_year)
  short_year = year[0:4]
  arr1 = np.append(arr1,short_year)

unique_years = set(arr1)
print("unique_years:")
print(unique_years)
print()

unique_arr = np.array(pd.Data frame.from_dict(unique_
years))
print("unique_arr:")
print(unique_arr)
print()
```

Listing B.17 initializes df with the contents of the CSV file multiple_
dates.csv and then displays its contents. The next portion of Listing B.17
initializes year_list with the dates column of df.

The next code block contains a loop that iterates through the elements
in year_list, extracts the first four characters (i.e., the year value), and
appends that substring to the NumPy array arr1. The final code block initializes the variable unique_arr as a Numpy array consisting of the unique years
in the dictionary unique_years. Launch the code in Listing B.17 to see the
following output:

```
df:
        dates   values
0   2020-01-31    40.0
1   2020-02-28    45.0
2   2020-03-31    56.0
3   2021-04-30     NaN
4   2021-05-31     NaN
5   2021-06-30   140.0
```

```
6   2021-07-31      95.0
7   2022-08-31      40.0
8   2022-09-30      55.0
9   2022-10-31       NaN
10  2022-11-15      65.0

unique_years:
{'2022', '2020', '2021'}

unique_arr:
[['2022']
 ['2020']
 ['2021']]
```

Listing B.18 shows the content of `pandas_misc2.py` for iterating through the rows of a data frame. Row-wise iteration is not recommended because it can result in performance issues in larger datasets.

Listing B.18: pandas_misc2.py

```
import pandas as pd

df = pd.read_csv('multiple_dates.csv', parse_
dates=['dates'])

print("df:")
print(df)
print()

print("=> ITERATE THROUGH THE ROWS:")
for idx,row in df.iterrows():
  print("idx:",idx," year:",row['dates'])
print()
```

Listing B.18 initializes the data frame `df`, prints its contents, and then processes the rows of `df` in a loop. During each iteration, the current index and row contents are displayed. Launch the code in Listing B.18 to see the following output:

```
df:
        dates   values
0   2020-01-31    40.0
1   2020-02-28    45.0
2   2020-03-31    56.0
3   2021-04-30     NaN
4   2021-05-31     NaN
5   2021-06-30   140.0
6   2021-07-31    95.0
7   2022-08-31    40.0
8   2022-09-30    55.0
9   2022-10-31     NaN
10  2022-11-15    65.0


=> ITERATE THROUGH THE ROWS:
idx: 0   year: 2020-01-31 00:00:00
idx: 1   year: 2020-02-28 00:00:00
idx: 2   year: 2020-03-31 00:00:00
idx: 3   year: 2021-04-30 00:00:00
idx: 4   year: 2021-05-31 00:00:00
idx: 5   year: 2021-06-30 00:00:00
idx: 6   year: 2021-07-31 00:00:00
idx: 7   year: 2022-08-31 00:00:00
idx: 8   year: 2022-09-30 00:00:00
idx: 9   year: 2022-10-31 00:00:00
idx: 10  year: 2022-11-15 00:00:00
```

Listing B.19 shows the content of `pandas_misc3.py` for displaying a weekly set of dates that are between a start date and an end date.

Listing B.19: pandas_misc3.py

```
import pandas as pd

start_d="01/02/2022"
end_d="12/02/2022"
weekly_dates=pd.date_range(start=start_d, end=end_d,
freq='W')

print("Weekly dates from",start_d,"to",end_d,":")
print(weekly_dates)
```

Listing B.19 starts with initializing the variable start_d and end_d that contain a start date and an end date, respectively, and then initializes the variable weekly_dates with a list of weekly dates between the start date and the end date. Launch the code in Listing B.19 to see the following output:

```
Weekly dates from 01/02/2022 to 12/02/2022 :
DatetimeIndex(['2022-01-02', '2022-01-09', '2022-01-
16', '2022-01-23',
                '2022-01-30', '2022-02-06', '2022-02-
13', '2022-02-20',
                '2022-02-27', '2022-03-06', '2022-03-
13', '2022-03-20',
                '2022-03-27', '2022-04-03', '2022-04-
10', '2022-04-17',
                '2022-04-24', '2022-05-01', '2022-05-
08', '2022-05-15',
                '2022-05-22', '2022-05-29', '2022-06-
05', '2022-06-12',
                '2022-06-19', '2022-06-26', '2022-07-
03', '2022-07-10',
                '2022-07-17', '2022-07-24', '2022-07-
31', '2022-08-07',
                '2022-08-14', '2022-08-21', '2022-08-
28', '2022-09-04',
                '2022-09-11', '2022-09-18', '2022-09-
25', '2022-10-02',
```

```
                '2022-10-09', '2022-10-16', '2022-10-
23', '2022-10-30',
                '2022-11-06', '2022-11-13', '2022-11-
20', '2022-11-27'],
               dtype='datetime64[ns]', freq='W-SUN')
```

MERGING AND SPLITTING COLUMNS IN PANDAS

Listing B.20 shows the contents of employees.csv, and Listing B.21 shows the contents of emp_merge_split.py. These examples illustrate how to merge columns and split columns of a CSV file.

Listing B.20: employees.csv

```
name,year,month
Jane-Smith,2015,Aug
Dave-Smith,2020,Jan
Jane-Jones,2018,Dec
Jane-Stone,2017,Feb
Dave-Stone,2014,Apr
Mark-Aster,,Oct
Jane-Jones,NaN,Jun
```

Listing B.21: emp_merge_split.py

```python
import pandas as pd

emps = pd.read_csv("employees.csv")
print("emps:")
print(emps)
print()

emps['year']  = emps['year'].astype(str)
emps['month'] = emps['month'].astype(str)
```

```
# separate column for first name and for last name:
emps['fname'],emps['lname'] = emps['name'].str.
split("-",1).str


# concatenate year and month with a "#" symbol:
emps['hdate1'] = emps['year'].
astype(str)+"#"+emps['month'].astype(str)


# concatenate year and month with a "-" symbol:
emps['hdate2'] = emps[['year','month']].agg('-'.join,
axis=1)


print(emps)
print()
```

Listing B.21 initializes the data frame df with the contents of the CSV file employees.csv, and then displays the contents of df. The next pair of code snippets invoke the astype() method to convert the contents of the year and month columns to strings.

The next code snippet in Listing B.21 uses the split() method to split the name column into the columns fname and lname that contain the first name and last name, respectively, of each employee's name:

```
emps['fname'],emps['lname'] = emps['name'].str.
split("-",1).str
```

The next code snippet concatenates the contents of the year and month string with a "#" character to create a new column called hdate1:

```
emps['hdate1'] = emps['year'].
astype(str)+"#"+emps['month'].astype(str)
```

The final code snippet concatenates the contents of the year and month string with a "-" to create a new column called hdate2, as shown here:

```
emps['hdate2'] = emps[['year','month']].agg('-'.join,
axis=1)
```

Launch the code in Listing B.21 to see the following output:

```
emps:
          name      year month
0  Jane-Smith  2015.0    Aug
```

```
1   Dave-Smith   2020.0    Jan
2   Jane-Jones   2018.0    Dec
3   Jane-Stone   2017.0    Feb
4   Dave-Stone   2014.0    Apr
5   Mark-Aster      NaN    Oct
6   Jane-Jones      NaN    Jun
```

```
          name    year month  fname   lname       hdate1
hdate2
0  Jane-Smith   2015.0    Aug   Jane   Smith   2015.0#Aug
2015.0-Aug
1  Dave-Smith   2020.0    Jan   Dave   Smith   2020.0#Jan
2020.0-Jan
2  Jane-Jones   2018.0    Dec   Jane   Jones   2018.0#Dec
2018.0-Dec
3  Jane-Stone   2017.0    Feb   Jane   Stone   2017.0#Feb
2017.0-Feb
4  Dave-Stone   2014.0    Apr   Dave   Stone   2014.0#Apr
2014.0-Apr
5  Mark-Aster      nan    Oct   Mark   Aster      nan#Oct
nan-Oct
6  Jane-Jones      nan    Jun   Jane   Jones      nan#Jun
nan-Jun
```

There is one other detail regarding the following commented-out code snippet:

```
#emps['fname'],emps['lname'] = emps['name'].str.
split("-",1).str
```

The following deprecation message is displayed if you uncomment the preceding code snippet:

```
#FutureWarning: Columnar iteration over characters
#will be deprecated in future releases.
```

READING HTML WEB PAGES

Listing B.22 displays the contents of the HTML Web page abc.html, and Listing B.23 shows the content of read_html_page.py that illustrates how to read the contents of an HTML Web page from Pandas. Note that this code will only work with Web pages that contain *at least* one HTML <table> element.

Listing B.22: abc.html

```
<html>
<head>
</head>
<body>
  <table>
    <tr>
      <td>hello from abc.html!</td>
    </tr>
  </table>
</body>
</html>
Listing B.23: read_html_page.py
import pandas as pd

file_name="abc.html"
with open(file_name, "r") as f:
  dfs = pd.read_html(f.read())

print("Contents of HTML Table(s) in the HTML Web
Page:")
print(dfs)
```

Listing B.23 starts with an import statement, followed by initializing the variable file_name to abc.html that is displayed in Listing B.22. The next code snippet initializes the variable dfs as a data frame with the contents of the HTML Web page abc.html. The final portion of Listing B.19 displays the contents of the data frame `dsf`. Launch the code in Listing B.23 to see the following output:

```
Contents of HTML Table(s) in the HTML Web Page:
[                          0
0  hello from abc.html!]
```

For more information about the Pandas `read_html()` method, navigate to this URL:

https://pandas.pydata.org/pandas-docs/stable/reference/api/pandas.
read_html.html

SAVING A PANDAS DATA FRAME AS AN HTML WEB PAGE

Listing B.24 shows the content of `read_html_page.py` that illustrates how to read the contents of an HTML Web page from Pandas. Note that this code will only work with Web pages that contain at least one HTML `<table>` element.

Listing B.24: read_html_page.py

```python
import pandas as pd

emps = pd.read_csv("employees.csv")
print("emps:")
print(emps)
print()

emps['year']  = emps['year'].astype(str)
emps['month'] = emps['month'].astype(str)

# separate column for first name and for last name:
emps['fname'],emps['lname'] = emps['name'].str.
split("-",1).str

# concatenate year and month with a "#" symbol:
emps['hdate1'] = emps['year'].
astype(str)+"#"+emps['month'].astype(str)
```

```
# concatenate year and month with a "-" symbol:
emps['hdate2'] = emps[['year','month']].agg('-'.join,
axis=1)

print(emps)
print()

html = emps.to_html()
print("Data frame as an HTML Web Page:")
print(html)
```

Listing B.24 populates the data frame `temps` with the contents of `employees.csv`, and then converts the `year` and `month` attributes to type string. The next code snippet splits the contents of the `name` field with the "-" symbol as a delimiter. As a result, this code snippet populates the new `fname` and `lname` fields with the first name and last name, respectively, of the previously split field.

The next code snippet in Listing B.24 converts the `year` and `month` fields to strings, and then concatenates them with a "#" as a delimiter. Yet another code snippet populates the `hdate2` field with the concatenation of the `year` and `month` fields.

After displaying the content of the data frame `emps`, the final code snippet populates the variable `html` with the result of converting the data frame `emps` to an HTML Web page by invoking the `to_html()` method of Pandas. Launch the code in Listing B.24 to see the following output:

```
Contents of HTML Table(s)
emps:
          name      year month
0   Jane-Smith    2015.0   Aug
1   Dave-Smith    2020.0   Jan
2   Jane-Jones    2018.0   Dec
3   Jane-Stone    2017.0   Feb
4   Dave-Stone    2014.0   Apr
5   Mark-Aster       NaN   Oct
6   Jane-Jones       NaN   Jun
```

```
         name     year month fname   lname      hdate1
hdate2
0  Jane-Smith  2015.0   Aug   Jane   Smith   2015.0#Aug
2015.0-Aug
1  Dave-Smith  2020.0   Jan   Dave   Smith   2020.0#Jan
2020.0-Jan
2  Jane-Jones  2018.0   Dec   Jane   Jones   2018.0#Dec
2018.0-Dec
3  Jane-Stone  2017.0   Feb   Jane   Stone   2017.0#Feb
2017.0-Feb
4  Dave-Stone  2014.0   Apr   Dave   Stone   2014.0#Apr
2014.0-Apr
5  Mark-Aster     nan   Oct   Mark   Aster      nan#Oct
nan-Oct
6  Jane-Jones     nan   Jun   Jane   Jones      nan#Jun
nan-Jun

Data frame as an HTML Web Page:
<table border="1" class="data frame">
  <thead>
    <tr style="text-align: right;">
      <th></th>
      <th>name</th>
      <th>year</th>
      <th>month</th>
      <th>fname</th>
      <th>lname</th>
      <th>hdate1</th>
      <th>hdate2</th>
    </tr>
  </thead>
  <tbody>
    <tr>
      <th>0</th>
      <td>Jane-Smith</td>
      <td>2015.0</td>
```

```
        <td>Aug</td>
        <td>Jane</td>
        <td>Smith</td>
        <td>2015.0#Aug</td>
        <td>2015.0-Aug</td>
      </tr>
      <tr>
        <th>1</th>
        <td>Dave-Smith</td>
        <td>2020.0</td>
        <td>Jan</td>
        <td>Dave</td>
        <td>Smith</td>
        <td>2020.0#Jan</td>
        <td>2020.0-Jan</td>
      </tr>
      // details omitted for brevity
      <tr>
        <th>6</th>
        <td>Jane-Jones</td>
        <td>nan</td>
        <td>Jun</td>
        <td>Jane</td>
        <td>Jones</td>
        <td>nan#Jun</td>
        <td>nan-Jun</td>
      </tr>
    </tbody>
  </table>
```

SUMMARY

This appendix introduced you to Pandas for creating labeled data frames and displaying the metadata of data frames. Then, you learned how to create data frames from various sources of data, such as random numbers and hard-coded data values. In addition, you saw how to perform column-based and row-based operations in Pandas data frames.

You also saw how to create data frames with various types of data, such as numeric and Boolean data frames. In addition, we discuss examples of creating data frames with NumPy functions and random numbers. Furthermore, you learned how to create Pandas data frames from data stored in CSV files.

INDEX

A

A/B testing, 161–165
  A/B/n testing, 164
  confidence level, 163
  errors, 163–164
  interleaving algorithm, 164
  multivariate testing (MVT), 165
  quasi-experiment, 165
  sequential, 162–163
  split test, 161
  statistical errors, 164
accuracy, 84
accuracy and balanced accuracy, 82,
  108–109
adjusted R², 98–99
advanced probability functions, 136–139
  beta distribution, 137–138
  zeta distribution, 138–139
alpha value, 154–155
  in conjunction with p-value, 157
alternate hypothesis (H1), 149–150
Anaconda `Python` distribution, 169
arithmetic mean, 61–62
artificial counterfactual estimation (ACE), 69
AUC (Area Under a Curve), 88–91
audio data, 6

B

bar charts, 3
Bernoulli distribution, 108–109
beta distribution, 137–138

bias, 72–73
binary confusion matrix, 80
binary data, 7, 12
"binning" data values, 15–16
  issues, 16
  programmatic binning techniques, 15–16
binomial distribution, 109–111, 140
black-box shift detector, 13
Boolean operations, 216
box plots, 3
Brier score, 96–97

C

card drawing probabilities, 47–49
cardinality, 29
categorical data, 8, 27, 56, 93–97
categorical data to numeric data conversion,
  `Pandas`, 222–227
Central Limit Theorem (CLT), 74, 124
Chebyshev's inequality, 64
children-related probabilities, 52–53
chi-squared distribution, 117–118
`chr()` function, 183
class imbalance, 72
cluster sampling, 71
Cohen's kappa coefficient, 83
coin tossing probabilities, 37–43
  solving "at least 2" and "at most (n-2)"
    tasks, 42–43
  solving "at least" and "at most" tasks, 41–42
  threshold, 39–41

columns, merging and splitting in `Pandas`, 244–246

command-line arguments, `Python`, 204–205

compile time and runtime code checking, 180

components, 151–152

conditional probability, 33

confidence interval (CI), 158–160

confidence level, 158, 161, 163

confirmation bias, 72

confounding variable, 68

confusion matrix, 75–87

container-based probabilities, 49–52
 complement, 50–51
 matching pair, selection, 51–52

continuous data
 vs. discrete data, 9–10
 metrics for, 97–100

continuous probability distributions, 116–136
 chi-squared distribution, 117–118
 exponential distribution, 118–120
 F-distribution, 120–122
 gamma distribution, 122–123
 Gaussian distribution, 124–127
 isotropic Gaussian distribution, 127
 lognormal distribution, 127–128
 lognormal graph, 128–129
 multiple lognormal graphs, 130–132
 t-distribution, 132–134
 uniform distribution, 134–136

continuous random variable, 55, 67–68, 106, 156

continuous variable, 14, 56, 122

control group, 162

convergence for random variables, 70

correlation, 17–19
 correlation matrix, 17
 discrimination threshold, 18
 `Python` code sample, 17

`corr()` method, 17

counterfactual analysis, 68–69

covariate drift, 13

CSV (comma-separated values) files, 2, 4–6, 9
 reading in `Pandas`, 220–222

cultural bias, 73

cumulative density function (CDF), 106

D

data
 accuracy, 3
 cleaning
  cost, 2
  and data wrangling steps, 4–5
 dealing with, 5–7
 drift, 1, 13
 frames and data cleaning tasks, `Pandas`, 209–210
 governance, 4
 leakage, 1
 literacy, 1–2
 management, 3
 quality, 1, 3–4
 stewardship, 4
 types
  explanation of, 7–11
  in, `Python`, 180–181
  working with, 12–13
 wrangling steps, 4–5

data-centric AI *vs.* model-centric AI, 1, 4

DataFrame, `Pandas`
 describing, 213–216
 as HTML Web page, 247–248
 with `NumPy` example, 210–213
 and random numbers, 218–220
 transposing, 217–218

datasets, 6–7

dates-related functions, 196–198

decision-making method, 163

degrees of freedom, 120

DeMorgan's laws, 30

descriptive statistics, 56

`df`, 23, 79

dice tossing probabilities, 43–47
  solving "at least 2" and "at most (n-2)"
    tasks, 44–45
  solving "at least" and "at most" tasks, 44
  solving unequal dice values, 45–46
  working with k-sided dice, 46–47
digits and alphabetic characters, testing for,
  191–192
dimensionality reduction algorithms, 6
discrete data *vs.* continuous data, 9–10, 14
discrete probability distributions
  Bernoulli distribution, 108–109
  binomial distribution, 109–111
  geometric distribution, 114–116
  Poisson distribution, 111–114
  Poisson values, generating, 112–113
discrete variable, 56
discrete *vs.* continuous random variables,
  56, 67–68
distfit fits, 140–147
distribution, 32
DMV (department of motor vehicles), 18
DoesNotExist, 180
domain expert, 2, 6
drift, 13
dynamic dataset, 6

E

emojis (optional) and Python, 203
empty set, 29
error matrix, 80
Excel spreadsheet, 6
exception handling in Python, 199
exploratory data analysis (EDA), 1–5
  data-centric AI or model-centric AI, 4
  data cleaning and data wrangling steps,
    4–5
  data quality, 3–4
  purpose, 2
  types, 3
  visualization techniques, 3
exponential distribution, 118–120, 140

F

F1-related scores, 96
F1 score, 93–94
Faker, 19–23
  launching Faker from command line,
    20–21
  Python code sample with, 20
false discovery rate (FDR), 86
false omission rate (FOR), 86
false positive rate (FPR), 87
F-distribution, 120–122
feature, drift, 13
formatting numbers in Python, 184–185
Fraction() function, 185
frequency, 32
frequency distribution, 56

G

gamma distribution, 122–123
Gaussian distribution, 124–127
geographical bias, 73
geometric distribution, 114–116
geometric mean, 61–62

H

handling user input, 200–202
harmonic mean, 61–62
HELP() and DIR() functions,
  178–180
heterogenous sources of data, 3
histograms, 3
housing prices, 6, 102
HTML Web page, Pandas, 247–248
hypothesis testing, 149
  A/B testing, 161–165
  alpha value, 154–155
    in conjunction with p-value, 157
  alternate hypothesis (H1), 149–150
  components, 151–152
  confidence interval (CI), 158–160
  confidence level, 158, 161

Maximum Likelihood Estimation (MLE), 166–167
non-parametric tests, 151
null hypothesis (H0), 149–150
parametric tests, 151
point estimation, 157
p-value, 154–155
significance level, 161
statistically significant results, 150–151
test statistics, 152–153
z-scores, 155–156

I

identifiers, 174
image data, 6
independent sample, 55
inferential statistics, 56
input drift, 13
installation, Python, 171–172
integer-based ordinal data, 8
integer-based values, 10
integrated development environment (IDE), 172–173
interleaving algorithm, 164
interquartile values, 59–60
interval data, 9
IPython tool, 170–171
isotropic Gaussian distribution, 127

J

Java, 174

K

k-sided dice, 46–47
kurtosis, 65–67

L

label bias, 72
labeling data, 3
largest-extreme-value distribution, 140

launching, Python, 172–173
Law of Large Numbers (LLN), 73–74
The Law of Large Numbers, 32
line
    graphs, 3
    indentation, and multilines, 174–175
linear data approximation with np.linspace(), 102–103
linear regression, 93–94, 97–98, 102
LOC() and ILOC() methods in Pandas, 222
lognormal distribution, 127–128, 140
lognormal graph, 128–129

M

machine learning
    classification algorithms, 76, 93
    model, 5
Mann-Whitney U test, 83
matching and splitting strings in Pandas, 227–230
Matthews correlation coefficient (CCM), 83
Maximum Likelihood Estimation (MLE), 166–167
mean, 57
    vs.median, 58–59
Mean Absolute Error (MAE), 100–102
Mean Squared Error (MSE), 100–102
measurement bias, 72
median, 58
media-related data, 11
mode, 59
model drift, 13
model performance bias, 73
module in Python, 177
multiple lognormal graphs, 130–132
multiple random variables, 69–70
multivariate charts, 3
multivariate testing (MVT), 165

N

negative predictive value (NPV), 85
no-code solution, 83
NOIR, statistical data, 11
nominal data, 7–8
non-Gaussian distributions, 140
non-linear least squares, 102
non-normal distribution, 140
non-parametric tests, 151
normalized confusion matrix, 77
NoSQL database, 6
null hypothesis (H0), 149–150
numeric data, 1, 6, 12, 78, 222–227

O

objective-C, 174
observation or measurement, 56
ordinal data, 8
outcome space, probability, 36

P

paired sample, 55
Pandas
    alternatives to, 210
    Boolean operations, 216–218
    categorical data to numeric data
        conversion, 222–227
    DataFrame, 209
        describing, 213–216
        as HTML Web page, 248–251
        with NumPy example, 210–213
        and random numbers, 218–220
        transposing, 217–218
    data frames and data cleaning tasks,
        209–210
    date ranges in, 232–234
    dates in, operations with, 239–244
    documentation, 7
    HTML Web page, 247–248
    LOC() and ILOC() methods in,
        222

matching and splitting strings in,
    227–230
merging and splitting columns in,
    244–246
missing dates
    detection in, 234–236
    interpolation in, 236–239
    options and settings, 208–209
    reading csv files in, 220–222
    strings to dates conversion in, 230–232
parametric tests, 151
PATH environment variable, 172
point estimation, 157
Poisson distribution, 111–114, 118, 140
population, 55–56
precision, 75, 84
prevalence, 84
Principal Component Analysis (PCA), 6
printing text without new line characters,
    194–195
probability
    card drawing probabilities, 47–49
    children-related probabilities, 52–53
    coin tossing probabilities, 37–43
    concepts in, 32–36
    container-based probabilities, 49–52
    dice tossing probabilities, 43–47
    of item selection
        binary case, 34
        multi-class case, 34–35
    joint probability, 35
    mutually exclusive events, 35–36
    open, closed, compact, and convex sets
        (optional), 31–32
    sampling, 71
    set theory, 29–31, 36–37
probability density function (PDF), 106
probability distribution, 33
    advanced probability functions, 136–139
    best-fitting distribution for data, 140–147
    continuous probability distributions,
        116–136

cumulative density function (CDF), 106
discrete probability distributions
    Bernoulli distribution, 108–109
    binomial distribution, 109–111
    geometric distribution, 114–116
    Poisson distribution, 111–114
    Poisson values, generating, 112–113
non-Gaussian distributions, 140
probability density function (PDF), 106
probability mass function (PMF),
    106–107
types, 107–108
probability mass function (PMF), 106–107
p-value, 154–155
Python
    code samples, 17, 20
        of confusion matrix, 78–80
    command-line arguments, 204–205
    compile time and runtime code
        checking, 180
    dates, working with, 196–198
    and emojis (optional), 203
    exception handling in, 198–200
    fractions, working with, 185–186
    handling user input, 200–202
    HELP() and DIR() functions, 178–180
    identifiers, 174
    installation, 171–172
    launching, 172–173
    lines, indentation, and multilines,
        174–175
    module in, 177
    numbers, working with, 181–185
    PATH environment variable, 172
    printing text without newline characters,
        194–195
    quotation and comments in, 175–176
    remove leading and trailing characters,
        193–194
    search and replace a string in other
        strings, 192–193
    simple data types in, 180–181
    slicing and splicing strings, 190–192
    standard modules in, 178
    strings, working with, 187–190
    text alignment, 195–196
    tools for, 169–171
    unicode and UTF-8, 186–187
    unicode, working with, 186–187
    uninitialized variables and the value
        none in, 180
Python-based open-source package,
    140–141
Python interactive interpreter, 173

Q

qualitative data analysis, 3
quantitative data analysis, 3
quasi-experiment, 165
quotation marks and comments in Python,
    175–176

R

randomized controlled trials, 68–69
random sampling, 71
random variables, 33, 55, 67–69, 106, 120,
    166
ratio data, 9
recall, 83–85
Receiving Operator Characteristic (ROC)
    curves, 87, 92
remove leading and trailing characters,
    193–194
residual sum of squares (RSS), 97–98
revenue generation, 162
Root Mean Squared Error (RMSE),
    100–102
round() function in Python, 184
R-squared, 97–100

S

sample space, 30–31
sampling bias, 72
sampling techniques for population, 70–71

scaling, data values, 1
scikit-learn documentation, 8
search and replace a string in other strings, 192–193
selection bias, 72
sensitivity, 84–85
sequential A/B testing, 162–163
set theory, 29–31, 36–37
significance level, 154, 161
skewed *vs.* uniform distributions, 140
skewness, 65–67
sklearn.metrics module (optional), 92–93
slicing and splicing strings, 190–192
specificity, 84
split test, 161
Standard Library modules, 178
standard modules, 178
static dataset, 6
statistical data, types, 11
statistical errors, 164
statistically significant results, 150–151
statistics
    arithmetic mean, 61
    code sample, 63–64
    confounding variable, 68
    confusion matrix, 75–87
    continuous random variable, 68
    counterfactual analysis, 68–69
    descriptive statistics, 56
    discrete *vs.* continuous random variables, 67–68
    frequency distribution, 56
    geometric mean, 61–62
    harmonic mean, 93–94
    inferential statistics, 56
    interquartile values, 59–60
    mean, 57
    mean vs.median, 58–59
    median, 58
    mode, 59
    multiple random variables, 69–70
    random variable, 67–69

sampling techniques in for population, 70–71
    variance and standard deviation, 62–64
    weighted mean, 57
stock prices, 6, 14
strata, 71
stratified sampling, 71
string-based values, 10
strings in `Python2`, 187–190
strings to dates conversion in `Pandas`, 198, 230–232
Student's t-test, 83
synthetic data, working with, 19–27
    customer data, generating and saving, 21–23
    `fake_purch_orders.csv`, 23–24
    `Faker`, 19–23
    generating purchase orders (optional), 24–27
    `gen_purch_orders.py`, 24–25
    launching `Faker` from command line, 20–21
    `Python` code sample with `Faker`, 24
systematic sampling, 71

T

t-distribution, 132–134
`test.py`, 204
test statistics, 152–153
text alignment, 195–196
text-based data, 6–7
    documents, 11
    types, 7
theoretical probability, 32
threshold probabilities, 39–41
time-based data from IoT devices, 6
time bias, 72
Titanic dataset, 1
tools for `Python`
    `easy_install and pip`, 169–170
    `IPython`, 170–171
    `virtualenv`, 170

total sum of squares (TSS), 97–98
true positive rate (TPR), 85, 87
t-score, 153
TSV (tab-separated values) files, 6
TypeError, 199
type I and type II errors, 81–82

U

unequal dice values, 45–46
unicode and UTF-8, 186–187
uniform distribution, 134–136
uninitialized variables and value none in
    Python, 180

V

variables, 55
    confounding variable, 68
    continuous random variable, 68
    continuous variable, 56
    convergence for random variables, 70
    discrete variable, 56

discrete *vs.* continuous random variables,
    56, 67–68
multiple random variables, 69–70
PATH environment variable, 172
random variables, 67–69
synthetic data, working with, 19–27
uninitialized variables and value none in
    Python, 180
variance and standard deviation, 62–64
video data, 6
virtualenv tool, 170

W

Weibull distribution, 140
weighted mean, 57

Z

zero value, 18–19
zeta distribution, 138–139
z-scores, 153, 155–156
z-test, 153

www.ingramcontent.com/pod-product-compliance
Lightning Source LLC
Chambersburg PA
CBHW061352210326
41598CB00035B/5964